Organic Synthesis Using Samarium Diiodide
A Practical Guide

Organic Synthesis Using Samarium Diiodide

A Practical Guide

David J. Procter
School of Chemistry, The University of Manchester, Manchester, UK

Robert A. Flowers, II
Department of Chemistry, Lehigh University, Bethlehem, PA, USA

Troels Skrydstrup
University of Aarhus, Aarhus, Denmark

RSC Publishing

ISBN: 978-1-84755-110-8

A catalogue record for this book is available from the British Library

Published by The Royal Society of Chemistry,
Thomas Graham House, Science Park, Milton Road,
Cambridge CB4 0WF, UK

Registered Charity Number 207890

For further information see our web site at www.rsc.org

Foreword

I was delighted to write a Foreword for this book because it deals with an old friend, samarium diiodide. This reagent has given us much excitement and many surprises.

In 1977 we published our first article on samarium diiodide (*A new preparation of some divalent lanthanide iodides and their usefulness in organic synthesis*, J.-L. Namy, P. Girard and H. B. Kagan, *Nouv. J. Chim.*, 1977, **1**, 5), followed 3 years later by a full paper (*Divalent lanthanide derivatives in organic synthesis – I. Mild preparation of SmI_2 and YbI_2 and their use as reducing or coupling agents*, P. Girard, J. L. Namy and H. B. Kagan, *J. Am. Chem. Soc.*, 1980, **102**, 2693).

How did this new research direction come about when our work at that time was mainly oriented towards asymmetric synthesis? In 1971, we were involved in some collaborative works on ^1H NMR spectroscopy using simple europium(III) chelates. An explosion of papers in that new area led me to change the research topic of my PhD student, Pierre Girard, who was engaged in research on europium(III) reagents. I decided to keep the keyword *lanthanide* for the thesis, but chose to reorient the work towards organic reactions. We first established an air oxidation system for the conversions of benzoins to benzils, catalysed by $Yb(NO_3)_3$. We then moved to the little known field of synthesis using divalent lanthanides, with the expectation that new reductions of organic compounds would be realised. Divalent europium salts are easy to prepare and to handle and it had been described that an aqueous solution of $EuCl_2$ could reduce isonicotinic acid to the corresponding aldehyde. Unfortunately, we were not able to extend this chemistry to other classes of compound. We hypothesised that more promising reducing agents could be found using other lanthanides. At that time, the following oxidation potentials (in water) were reported (Ln^{3+}/Ln^{2+}): Eu –0.33, Yb –1.15, Sm –1.55, Tm –2.55 V. Since thulium was very expensive, only ytterbium and samarium were considered. We decided to prepare inorganic salts of Yb(II) and Sm(II) and succeeded in obtaining THF solutions of SmI_2 and YbI_2 by treating samarium or ytterbium powder with 1,2-diiodoethane. The promising reducing properties of samarium diiodide encouraged us to concentrate on this new reagent.

Although samarium diiodide was first reported in 1906, for many years it was only of interest for inorganic materials science. Its application in organic chemistry had never been studied. We recognised that samarium diiodide had the potential to act as a reducing agent (a one-electron donor) since Sm^{3+} is the stable oxidation state. Moreover, the solubility of SmI_2 in THF (~0.1 M) and change of colour from deep blue–green (Sm^{2+}) to yellow (Sm^{3+}) allowed us to screen various organic transformations readily. In our 1977 paper, we reported the possibility of using samarium diiodide to perform samarium Barbier reactions, selective reduction of aldehydes in the presence of ketones and Meerwein–Ponndorf–Verley (MPV) reductions. Our 1980 full paper included many more results obtained by P. Girard during his PhD work. Dr J.-L. Namy was also involved in our early studies on samarium diiodide chemistry and remained in my laboratory. In subsequent years, we developed various aspects of the reactions induced by the new reagent. Much of our work is summarised in my 2003 review article (*Twenty-five years of organic chemistry with diiodosamarium: an overview*, H. B. Kagan, *Tetrahedron*, 2003, **59**, 10351). Since the mid-1980s, groups all over the world have explored the scope of this reagent and published many unexpected and interesting results. The scope of the transformations induced by SmI_2 quickly enlarged and many reviews are now devoted to the reagent. Some of the reactions of SmI_2 are surprisingly rapid under mild conditions, allowing them to be used selectively in polyfunctional systems. The influence of simple additives (water, alcohols, amines, HMPA, *etc.*) can modulate the reactivity of SmI_2 and induce specific reactions, as reviewed in our 1999 review (*Influence of additives on the organic chemistry mediated by diiodosamarium*, H. B. Kagan and J. L. Namy, in *Topics in Organometallic Chemistry, Lanthanides: Chemistry and Use in Organic Synthesis*, ed. S. Kobayashi, Springer, Berlin, 1999, p. 155). Interestingly, many carbon–carbon bond-forming reactions mediated by SmI_2, involving polyfunctional molecules, can be achieved with a high degree of diastereoselectivity, due to chelating effects involving Sm^{2+} or Sm^{3+} ions.

The SmI_2 technologies developed during the past 30 years have become an important tool in organic synthesis. I am pleased to see a new book devoted exclusively to organic synthesis using samarium diiodide and that focuses on the practical aspects of using the reagent. I congratulate the authors on this enterprise. This book will be very useful for newcomers to the field and I hope it will stimulate the development of new chemistry using the reagent in exciting new areas of synthesis.

Professor Henri B. Kagan

Preface

Samarium diiodide (SmI_2) is one of the most important reducing agents available to the synthetic organic chemist. Although SmI_2 is used extensively, there is still a view that the reagent can be difficult to prepare and to use. In addition, SmI_2 can mediate such a wide variety of organic chemistry that potential new users are often overawed by the extensive primary (and secondary!) literature on the reagent. This practical guide seeks to provide an introduction to the reagent and the organic reactions that it can be used to carry out. Where possible, results obtained using the reagent are compared with those obtained from the use of alternative reagents. The reader will see that this is often impossible, as alternative reagent systems do not exist for many of the transformations that SmI_2 can bring about. The guide covers work from Kagan's seminal papers through to examples taken from the recent literature (references are cited using group leaders' names rather than the first author of the reference). Where relevant, representative experimental procedures are given so the reader can see how straightforward the reagent is to use in the laboratory. Rather than being a comprehensive review of the chemistry of SmI_2, we have attempted to adopt an 'an all you need to know' approach to the subject that nevertheless gives sufficient detail to satisfy the more curious reader.

David J. Procter
Robert A. Flowers, II
Troels Skrydstrup

Organic Synthesis Using Samarium Diiodide: A Practical Guide
By David J. Procter, Robert A. Flowers, II and Troels Skrydstrup

Published by the Royal Society of Chemistry, www.rsc.org

Acknowledgements

The authors would like to thank the following people for their help in the preparation of this book. In Manchester: Dr Lorna Duffy, Dr Matthew Helm, Dr Susannah Coote, Hassan Harb, Thomas Baker, Johannes Vogel, Karl Collins, Dixit Parmar and Madeliene Da Silva. In Bethlehem: Dr. Rebecca Miller, Dr. Dhandapani Sadasivam, Kimberly Tachovsky. In Aarhus: Dr Karl B. Lindsay, Dr Dacil Hernández Mesa, Rolf H. Taaning. In addition, the authors would like to acknowledge Hassan Harb's assistance with the preparation of the cover art and Lise Procter for advice on the cover design.

Contents

Organic Synthesis Using Samarium Diiodide: A Practical Guide
By David J. Procter, Robert A. Flowers, II and Troels Skrydstrup

Published by the Royal Society of Chemistry, www.rsc.org

Abbreviations

Å	Ångstrom
Ac	acetyl, –C(O)CH_3
AIBN	2,2′-azobis(2-methylpropionitrile)
aq.	aqueous
BINOL	1,1′-bi(2-naphthol)
Bn	benzyl
Boc	*tert*-butoxycarbonyl
Bu	butyl
Bz	benzoyl
c	cyclo
cat.	catalyst
Cbz	benzyloxycarbonyl
Cp	cyclopentadienyl
Cp*	pentamethylcyclopentadienyl
Cy	cyclohexyl
DBM	dibenzoylmethido
DBU	1,8-diazabicyclo[5.4.0]undec-7-ene
DET	diethyl tartrate
DMA	*N*,*N*-dimethylacetamide
DMAE	*N*,*N*-dimethylaminoethan-2-ol
DMAP	4-dimethylaminopyridine
DME	dimethoxyethane
DMP	Dess–Martin periodinane
DMPU	1,3-dimethyl-3,4,5,6-tetrahydro-2(1*H*)-pyrimidinone
dr	diastereoisomeric ratio
E	undefined electrophile
E	entgegen
E°	standard reduction potential
EDCI	1-ethyl-3-(3-dimethylaminopropyl)carbodiimide hydrochloride
ee	enantiomeric excess
equiv	equivalent
Et	ethyl
Fmoc	9-fluorenylmethyloxycarbonyl
FSPE	fluorous solid-phase extraction
GC	gas chromatography

GLC	gas–liquid chromatography
h	hour
Hal	undefined halogen
Hex	hexyl
HFIP	1,1,1,3,3,3-hexafluoroisopropanol
HMPA	hexamethylphosphoramide
HOBt	hydroxybenzotriazole
HPLC	high-performance liquid chromatography
i	iso
k	rate constant
k_{cyc}	rate constant for cyclisation
k_{red}	rate constant for reduction
L	undefined ligand
LUMO	lowest unoccupied molecular orbital
M	molar
M	undefined metal
Me	methyl
MEM	2-methoxyethoxymethyl
min	minute
MOM	methoxy methyl ether
Ms	methanesulfonyl
MS	molecular sieves
MTM	methylthiomethyl
n	normal
nm	nanometre
NMR	nuclear magnetic resonance
Oct	octyl
Ph	phenyl
PMB	4-methoxybenzyl
PMDTA	*N*,*N*,*N*′,*N*′,*N*′-pentamethyldiethylenetriamine
PMP	4-methoxylphenyl
PPTS	pyridinium *p*-toluenesulfonate
Pr	propyl
RAMP	(*R*)-(+)-1-amino-2-(methoxymethyl)pyrrolidine
rt	room temperature
s	second
s	secondary
S	undefined solvent
SAMP	(*S*)-(–)-1-amino-2-(methoxymethyl)pyrrolidine
t	tertiary
TBAF	tetra-*n*-butylammonium fluoride
TBDMS	*tert*-butyldimethylsilyl
TBDPS	*tert*-butyldiphenylsilyl
tert	tertiary
TES	triethylsilyl
Tf	trifluoromethanesulfonyl

TFA	trifluoroacetic acid
THF	tetrahydrofuran
THP	tetrahydropyran
TIPS	triisopropylsilyl
TLC	thin-layer chromatography
TMEDA	*N*,*N*,*N*′,*N*′-tetramethylethylenediamine
TMS	trimethylsilyl
TMU	1,1,3,3-tetramethylurea
$Tp^{Me}{}_2$	hydrotris(3,5-dimethylpyrazolyl)borate
Trt	trityl (triphenylmethyl)
Ts	4-methylbenzenesulfonyl
Val	valine
W	watt
X	undefined heteroatom
Z	zussammen

CHAPTER 1

Introduction

1.1 Organic Synthesis Using Samarium Diiodide: A Practical Guide

1.1.1 Aims of the Book

Since its introduction to the synthetic chemistry community in 1977 by Kagan, samarium diiodide (SmI_2) has captured the imagination of organic chemists and has become one of the most important reducing agents available in the laboratory. The main chapters of this practical guide deal with the remarkable ability of SmI_2 to transform functional groups selectively and to orchestrate carbon–carbon bond formation. Other chapters deal with our understanding of mechanism and additive effects in reactions mediated by SmI_2 – an area that should still be considered as very much a work in progress. The final chapter of the book deals with selected emerging areas in the use of SmI_2 in synthesis and reflects the authors' research interests.

The book aims to steer a difficult course by providing a sufficient level of detail and new developments without burying the basics. *Representative procedures* have been included to encourage the reader to take their first steps into the fascinating organic chemistry of SmI_2.

1.1.2 Further Reading

The many excellent reviews on the use of SmI_2 in organic chemistry are a rich source of additional information. The major reviews and their authors are categorised according to their coverage below in Figure 1.1.

1.2 Introducing the Reagent

1.2.1 Working with SmI_2

Samarium(II) iodide (SmI_2) is commercially available as a solution in THF or can be prepared readily using one of several straightforward methods that have

Organic Synthesis Using Samarium Diiodide: A Practical Guide
By David J. Procter, Robert A. Flowers, II and Troels Skrydstrup

Published by the Royal Society of Chemistry, www.rsc.org

General reviews on SmI_2 chemistry

Soderquist, J. A., *Aldrichimica Acta,* **1991**, *24*, 15.
Molander, G.A., *Chem. Rev.*, **1992**, *92*, 29.
Molander, G.A., *Org. React.*, **1994**, *46*, 211.
Kagan, H. B., *Tetrahedron,* **2003**, *59*, 10351.

Reviews on sequential reaction using SmI_2

Molander, G.A.; Harris, C. R., *Chem. Rev.*, **1996**, *96*, 307.
Skrydstrup,T., *Angew. Chem. Int. Ed. Engl.*, **1997**, *36*, 345.
Molander, G. A.; Harris, C. R., *Tetrahedron,* **1998**, *54*, 3321.

The use of additive and co-solvent effects in SmI_2 chemistry

Kagan, H. B.; Namy, J.–L., *Lanthanides: Chemistry and Use in Organic Synthesis*; S. Kobayashi, Ed.; Springer, 1999, 155.
Dahlén, G. Hilmersson, *Eur. J. Inorg. Chem.*, **2004**, 3393.
Flowers, R. A. II, *Synlett.*, **2008**, 1427.

Specialist reviews on SmI_2

Barbier reactions – Krief, A.; Laval, A–M., *Chem. Rev.*, **1999**, *99*, 745.
Polymer chemistry – Nomura, R.; Endo, T., *Chem. Eur. J.*, **1998**, *4*, 1605.
Polymer chemistry – Agarwal, S.; Greiner, A. *J. Chem. Soc., Perkin Trans. 1*, **2002**, 2033.
Cyclisations in natural product synthesis – Edmonds, D. J.; Johnston, D.; Procter, D. J.; *Chem. Rev.*, **2004**, *104*, 3371.
β-Elimination reactions – Concellón, J. M.; Rodríguez – Solla, H., *Chem. Soc. Rev.*, **2004**, *33*, 599.
Results from the Kim group – Jung, D. Y.; Kim, Y. H., *Synlett.*, **2005**, 3019.
Applications in a symmetric synthesis – Gopalaiah, K.; Kagan, H. B., *New. J. Chem.*, **2008**, *32*, 607.
The chemistry of samarium enolates – Rudkin, I. M.; Miller,L. C.; Procter, D. J., *Organomet. Chem.*, **2008**, *34*, 19.

Other useful reviews that cover SmI_2 chemistry

Imamoto, T., *Lanthanides in Organic Synthesis*, Best Synthetic Methods, Academic Press, **1994**.
Gansauer, A.; Bluhm, H., *Chem. Rev.*, **2000**, *100*, 2771.
Steel, P. G., *J. Chem. Soc., Perkin Trans. 1*, **2001**, 2727.
Flowers, R. A., II; Prasad, E., in *Handbook on the Physics and Chemistry of Rare Earths*, Eds. K. A. Gschneidner, Jr. J.-C,G, Bunzli, V.K. Pecharsky, Elsevier, **2006**, Vol. 36, Chapter 230, pg. 393.

Figure 1.1 Reviews on samarium diiodide.

been described (see Chapter 2, Section 2.1). SmI_2 is air sensitive, but is tolerant of water and can be handled using standard syringe techniques. Reactions are typically carried out in THF although the use of other solvents has been investigated. Part of the reagent's popularity arises from its ability to mediate both radical and anionic processes and sequences involving both. As a result, it has been utilised in a wide range of synthetic transformations ranging from functional group interconversions to carbon–carbon bond-forming reactions. In addition, the reagent is often highly chemoselective and transformations instigated by SmI_2 tend to proceed with high degrees of stereoselectivity. Further adding to its appeal, the reactivity, chemoselectivity and stereoselectivity of SmI_2 can be manipulated and fine-tuned by the addition of various salts and cosolvents to the reaction mixture (see Chapter 2, Section 2.2).

1.2.2 Electronic Configuration of Sm(II)

Samarium, like all lanthanide elements, preferentially exists in the +3 oxidation state. The loss of the three outermost electrons, namely the $5d^1$, $6s^2$ electrons, results in enhanced thermodynamic stability in which a closed-shell

Xe-like electronic configuration is adopted. The +2 oxidation state is most relevant for samarium (f^6, near half-filled), europium (f^7, half-filled), thulium (f^{13}, nearly filled) and ytterbium (f^{14}, filled). In order to attain the more stable +3 oxidation state, SmI_2 readily gives up its final outer-shell electron, in a thermodynamically driven process, making it a very powerful and synthetically useful single-electron transfer reagent.

1.2.3 Reduction Potential

The redox potential of the SmI_2–SmI_2^+ couple has been established through the use of linear sweep and cyclic voltammetry and was found to be approximately −1.41 V, determined for a solution of SmI_2 in THF.[1] Although the preparation of SmI_2 in a variety of different solvents is known, THF is by far the most common solvent associated with its use. It would be expected, however, that the observed reduction potential of the reagent will vary greatly depending on the choice of solvent in which the measurement is carried out owing to the differences in the strength and number of solvent interactions.

It is possible to manipulate the reduction potential of SmI_2 through the use of various additives; most commonly these are found to be molecules containing neutral or Lewis basic oxygen functionalities. The most common example of this is the use of HMPA (hexamethylphosphoramide) as an additive, for which it was determined that, upon addition of 4 equiv, the reduction potential of SmI_2 in THF is increased to approximately −1.79 V, thereby significantly increasing its potency as a single-electron transfer reagent.[1a,2] Similar effects are also observed in the presence of alcohols and even water, for which it has been reported that upon the inclusion of 500 equiv, the reduction potential can be increased as far as −1.9 V.[3] The use and mechanistic role of such additives will be discussed in more detail in Chapter 2, Section 2.2.

1.2.4 Coordination Chemistry

Lanthanides typically adopt coordination numbers greater than six, depending largely on the size of the lanthanide ion and on the size of the ligands. SmI_2 is an oxophilic reagent and, as such, much of the coordination chemistry observed involves the close association of oxygenated molecules, including solvents and substrates, to the metal centre.

It has been established that in a solution of THF, SmI_2 exists in a *hepta*-coordinate geometry in which five THF molecules are equatorially bound to the central Sm(II) ion through their oxygen lone pairs, with iodide ligands axial.[4] The coordination number and geometry of SmI_2 is variable depending on the nature of the ligands involved. For example, in the $[SmI_2(HMPA)_4]$ complex alluded to previously, the more sterically demanding HMPA ligands are positioned equatorially around the Sm(II) ion, but this time only four ligands are involved and an octahedral (hexacoordinate) geometry results.[5,6] Complexes in which the coordination number is as high as eight

($[SmI_2\{O(CH_2CH_2OMe)_2\}_2]$) and nine ($[Sm\{O(CH_2CH_2OH)_2\}_3]I_2$) have also been identified.[7,8]

The oxophilic nature of SmI_2 is in many cases a highly beneficial quality: The coordination of two or more oxygenated reactive centres to samarium in both radical and ionic processes mediated by the reagent can often lead to high levels of diastereoselectivity in the formation of products. The coordination of Sm(II) and Sm(III) to oxygen donors on the substrate, solvent or cosolvent is a common theme that runs through each of the subsequent chapters.

References

1. (a) M. Shabangi and R. A. Flowers II, *Tetrahedron Lett.*, 1997, **38**, 1137; (b) R. J. Enemaerke, K. Daasbjerg and T. Skrydstrup, *Chem. Commun.*, 1999, 343.
2. (a) M. Shabangi, M. L. Kuhlman and R. A. Flowers II, *Org. Lett.*, 1999, 2133; (b) R. J. Enemaerke, T. Hertz, T. Skrydstrup and K. Daasbjerg, *Chem. Eur. J.*, 2000, **6**(20), 3747.
3. E. Prasad and R. A. Flowers II, *J. Am. Chem. Soc.*, 2005, **127**, 18093.
4. W. J. Evans, T. S. Gummersheimer and J. W. Ziller, *J. Am. Chem. Soc.*, 1995, **117**(35), 8999.
5. Z. Hou and Y. Wakatsuki, *J. Chem. Soc. Chem. Commun.*, 1994, 1205.
6. Z. Hou, Y. Zhang and Y. Wakatsuki, *Bull. Chem. Soc. Jpn.*, 1997, **70**, 149.
7. V. Chebolu, R. R. Whittle and A. Sen, *Inorg. Chem.*, 1985, **24**(19), 3082.
8. J. A. Teprovich Jr., M. N. Balili, T. Pintauer and R. A. Flowers II, *Angew. Chem. Int. Ed.*, 2007, **46**, 8160.

CHAPTER 2

The Reagent and the Effect of Additives

2.1 Preparing SmI_2

Although SmI_2 is commercially available as a 0.1 M solution in THF from several suppliers, it is also easy to prepare using one of several known procedures.

During Kagan's early work with SmI_2,[1] he developed a convenient method to prepare the reagent from samarium metal using 1,2-diiodoethane in THF (Scheme 2.1). Stirring this mixture under an inert atmosphere for several hours gave a 0.1–0.05 M solution of SmI_2 as a characteristic dark blue solution with the formation of ethene as a by-product. This solution is stable for several days when stored under an inert atmosphere, particularly when a small amount of samarium metal is present in the solution. Since Kagan's early studies, diiodomethane has been used by many groups, including that of Molander,[2] to oxidise samarium metal in place of 1,2-diiodoethane.

> *Representative procedure – preparation of SmI_2 using Kagan's procedure.* Samarium powder (3.00 g, 0.02 mol) was placed in a reaction flask under an inert atmosphere and a thoroughly degassed solution of 1,2-diiodoethane (2.82 g, 0.01 mol) in dry THF (250 ml) was slowly added. The mixture was stirred at room temperature until a dark blue SmI_2 solution was obtained.

Several years after Kagan published his preliminary work on SmI_2, Imamoto reported a more atom–efficient method for preparing SmI_2 that used samarium metal and iodine in THF.[3] Imamoto assumed that Kagan's route to the reagent proceeds by initial formation of samarium triiodide (SmI_3), which is then further reduced by samarium metal to give SmI_2 by a disproportionation process (Scheme 2.2).

> *Representative procedure – preparation of SmI_2 using Imamoto's procedure.* Samarium powder (1.00 g, 6.65 mmol) was placed in a reaction flask under an inert atmosphere and thoroughly degassed THF (55 ml) was added. Iodine (1.41 g,

Organic Synthesis Using Samarium Diiodide: A Practical Guide
By David J. Procter, Robert A. Flowers, II and Troels Skrydstrup

Published by the Royal Society of Chemistry, www.rsc.org

$$\mathrm{Sm} + \mathrm{ICH_2CH_2I} \xrightarrow{\text{THF, rt}} \boxed{\mathrm{SmI_2}} + \mathrm{H_2C{=}CH_2}$$

Scheme 2.1

Kagan's method using 1,2–diiodoethane

$$3\,\mathrm{Sm} + 3\,\mathrm{ICH_2CH_2I} \longrightarrow 2\,\mathrm{SmI_3} + 3\,\mathrm{CH_2{=}CH_2} + \mathrm{Sm}$$

$$2\,\mathrm{SmI_3} + \mathrm{Sm} \longrightarrow 3\,\mathrm{SmI_2}$$

Imamoto's method using iodine

$$2\,\mathrm{Sm} + 3\,\mathrm{I_2} \longrightarrow 2\,\mathrm{SmI_3} + \mathrm{Sm}$$

$$2\,\mathrm{SmI_3} + \mathrm{Sm} \longrightarrow 3\,\mathrm{SmI_2}$$

Scheme 2.2

$$\mathrm{Sm} + \underset{(\mathrm{CHI_3,\ CH_2I_2,\ 1,2\text{–}ICH_2CH_2I,\ I_2})}{\text{oxidant}} \xrightarrow[\text{or 180 °C, 5 min, MW, THF}]{\text{"sonication" THF, 5 min}} \mathrm{SmI_2}$$

Scheme 2.3

5.56 mmol) was added to the reaction mixture and the resulting suspension heated at 60 °C for 12 h to give a dark blue solution of SmI_2.

In 1992, Ishii reported the generation of a SmI_2 equivalent that displayed similar reactivity to SmI_2 by treatment of samarium metal with TMSCl and NaI.[4]

In recent years, exposing the preparation of the reagent from samarium metal and an oxidant to different stimuli has led to significant improvements in reaction time. Concellón utilized the sonication of samarium metal and iodoform at room temperature to give a solution of SmI_2 in THF in approximately 5 min (Scheme 2.3).[5] This approach was also used by Flowers to synthesise other Sm(II) species.[6] It was also reported that using different oxidants, such as 1,2-diiodoethane, diiodomethane and iodine, works just as well with this technique.[5]

Hilmersson[7] achieved similar results by heating the reaction to 180 °C under microwave conditions for 5 min (Scheme 2.3). Although this method is compatible with all of the oxidants described above, formation of SmI_2 from samarium metal and iodine is preferred as no gas evolution accompanies the reaction.[7]

As THF can act as a hydrogen donor towards radical species and can in some cases undergo ring opening in the presence of Lewis acidic samarium species, the solvent is not ideal for all SmI_2-mediated transformations.[8] This led various groups to investigate the formation of SmI_2 in other solvents. For example, Kagan and Namy reported the formation of the reagent in tetrahydropyran (THP),[8] whereas Ruder reported the use of acetonitrile,[9] and Tani prepared a solution in benzene and HMPA.[10] Kagan and Namy showed that the use of SmI_2 in THP allowed several transformations to be carried out that were not possible using the reagent in THF. For example, they succeeded in efficiently coupling acid chlorides, a transformation that was not possible using SmI_2 in THF,[8] and were able to form allylic and benzylic organosamariums, organometallic species that underwent decomposition in THF.[11,12] More recently, Flowers utilised sonication to prepare SmI_2 in low concentration (0.02–0.05 M) in acetonitrile, DME, 2-propanol, 2-methyl-2-propanol and 2-heptanol.[6] Attempts to prepare samarium diiodide in ethereal solutions other than THF, including diethyl ether, *tert*-butyl methyl ether and dioxane, were unsuccessful.[8]

2.2 The Use of Additives and Cosolvents in SmI_2 Reactions

One of the most fascinating features of SmI_2 is the ability to modify its behaviour through the use of cosolvents or additives. For example, cosolvents or additives can be used to control the rate of reduction or the chemo- or stereoselectivity of reactions. Additives commonly utilised to fine-tune the reactivity of SmI_2 can be classified into three major groups:

1. *Lewis bases* – HMPA and other electron-donor ligands, chelating ethers, *etc.*
2. *Proton sources* – predominantly alcohols and water.
3. *Inorganic additives* – NiI_2, $FeCl_3$, *etc.*

This section will focus on reactions that exemplify the use of particular additives and cosolvents and on the origin of the beneficial effects of the additives. Miscellaneous additives, not included in the three categories above, that promote the reactivity of SmI_2 will be discussed at the end of the section. Further examples of the use of additives and cosolvents can be found in subsequent chapters.

2.2.1 Lewis Bases

Lewis bases, containing basic nitrogen or oxygen atoms, are important promoters of reactions mediated by SmI_2. Among these additives HMPA has played an important role in the development of SmI_2-mediated reactions since it accelerates a wide range of functional group conversions and bond-forming reactions. The use of HMPA in SmI_2-mediated reductions not only increases the rate of reactions, but can also enhance the level of stereochemical control observed. Although numerous other Lewis basic additives have been used and alternative protocols have been developed, none yet approach the general utility of the SmI_2–HMPA reagent system. Thus, in spite of the toxicity of HMPA, it remains the additive of choice for many reactions using SmI_2. Illustrative examples of the use of HMPA as an additive are given below, and more detailed examples can be found in subsequent chapters.

The potential of the SmI_2–HMPA pairing was first recognized by Inanaga, who discovered that the use of this additive significantly increased the rate of reduction of alkyl and aryl halides.[13] The seminal work of Kagan showed that alkyl iodides are reduced by SmI_2 at elevated temperatures but alkyl bromides are reduced at a very slow rate and alkyl chlorides are unreactive.[1] Inanaga found that the addition of approximately 10% HMPA in THF dramatically enhanced the rate of reduction of alkyl halides: alkyl iodides and bromides were reduced at room temperature and even alkyl chlorides were reduced, albeit at elevated temperatures (Scheme 2.4). Synthetically relevant examples of alkyl halide reductions are described in Chapter 4, Section 4.2.

Inanaga showed that the presence of HMPA also accelerates the Barbier addition of alkyl halides to ketones and significantly improves the yield of the adducts.[14] He also found that HMPA was a useful additive in the SmI_2-mediated synthesis of lactones from bromo esters and ketones (Scheme 2.5). Since Inanaga's pioneering work, the Barbier reaction employing SmI_2–HMPA

$$\text{R–X} \xrightarrow{\text{SmI}_2\text{, THF}} \text{R–H}$$

R	X	reaction time	yield (%)
$C_{12}H_{25}$	I	6 h[a]	95
$C_{12}H_{25}$	Br	2 days[a]	82
$C_{12}H_{25}$	Cl	2 days[a]	no rxn

[a] room temperature [b] 60 °C

$$\text{R–X} \xrightarrow[\text{2-propanol}]{\text{SmI}_2\text{, THF–HMPA}} \text{R–H}$$

R	X	reaction time	yield (%)
$C_{10}H_{21}$	I	5 min[a]	> 95
$C_{10}H_{21}$	Br	10 min[a]	> 95
$C_{10}H_{21}$	Cl	8 h[b]	> 95

[a] room temperature [b] 60 °C

Scheme 2.4

R–Br → (SmI_2, THF; ketone)

R	Additive	reaction time	yield (%)
n-Bu	none	1 day[a]	67
n-Bu	HMPA	1 min	92
s-Bu	none	1.5 days[a]	27
s-Bu	HMPA	1 min	90

[a] reflux

Ph ketone, SmI_2, THF

R	Additive	reaction time	n	yield (%)
Me	none	120 min	2	39
Me	HMPA	1 min	2	85
Et	HMPA	1 min	3	55

Scheme 2.5

has been employed widely in synthesis and is discussed in detail in Chapter 5, Section 5.4.[15]

Although many early synthetic studies employed HMPA as a cosolvent, its mechanistic role remained unclear. Its role was later clarified by Molander, who studied the influence of HMPA concentration on the product distributions from the SmI_2-mediated reductive cyclisations of unactivated olefinic ketones.[16] The addition of HMPA was required to promote efficient ketyl-alkene cyclisation, and correlations between the concentration of HMPA, product ratios and diastereoselectivities were apparent (Scheme 2.6). In the absence of HMPA, attempted cyclisations led to the recovery of starting material **1**, reduced side-product **3** and desired cyclisation product **2**. Addition of 2 equiv of HMPA provided **2** and only a small fraction of **3**. Further addition of HMPA (3–8 equiv) provided **2** exclusively (Scheme 2.6).

The first clear trend in the data was that HMPA effectively increased the reducing ability of SmI_2. The absence of HMPA, or the presence of only small amounts (2 equiv), led to prolonged reaction times, while addition of a further 4–8 equiv of HMPA shortened reaction times. Molander suggested that HMPA may enhance the reducing ability in two ways: first, HMPA may dissociate

1. 2.2 SmI_2, THF, additive
2. $H_3O^{\oplus}$

conditions[a]	product (%)	time	%de (for **2**)	combined yield (%)
8 equiv HMPA	**2** (100)	<15 min	>99	90
4 equiv HMPA	**2** (100)	<15 min	>99	89
2 equiv HMPA	**2** (98), **3** (2)	2 h	96	91
none	**1** (33), **2** (62), **3** (5)	36 h	92	95

Scheme 2.6

SmI_2 aggregates in THF, making the reductant more reactive, second, HMPA may perturb the electron-donating orbital of Sm(II) and raise its energy, thus increasing the Sm(II)/Sm(III) reduction potential.[16] The second obvious trend in the data was the impact of HMPA concentration on product distribution. In the absence of HMPA, starting material **1** was recovered in addition to the desired product **2** and the side product **3**. The addition of 2 equiv of HMPA led to a significant increase in the yield of **2**. Molander also noted that **2** was obtained with improved diastereoselectivity on addition of HMPA. The addition of >4 equiv of HMPA provided **2** exclusively in high yield and with high diastereoselectivity. Based on these observations, Molander proposed that the addition of HMPA to SmI_2 produces a sterically encumbered reductant that not only enhances the diastereoselectivity of reactions, but also stabilises reactive intermediates (ketyls, radicals) in close proximity to Sm–HMPA complexes, thereby preventing competing reaction processes such as hydrogen atom abstraction from THF that result in the formation of side product **3**.

In the ensuing years, experimental studies have explored the hypotheses presented above. Crystallographic data reported by Evans provided evidence that the solvated structure of SmI_2 in THF is $[SmI_2(THF)_5]$.[17] Studies by Flowers provided solution evidence that deaggregation upon addition of HMPA is unimportant since vapour pressure osmometry experiments showed that SmI_2 existed as a monomer in THF and other electron donor solvents.[18] Initial insight into the structural details of the SmI_2–HMPA species was provided through crystallographic data obtained by Hou.[19] These studies showed that solutions containing 4 equiv of HMPA produced crystals of $[SmI_2(HMPA)_4]$ whereas solutions containing ≥ 10 equiv of HMPA produced crystals of octahedral $[Sm(HMPA)_6]I_2$.[20] Electrochemical studies by Flowers showed that the addition of HMPA creates a thermodynamically more powerful reductant.[21] Follow-up studies by Skrydstrup and Daasbjerg provided evidence that the addition of 4 equiv of HMPA to SmI_2 in THF results in the formation of $[Sm(THF)_2(HMPA)_4]I_2$, whereas the addition of ≥ 10 equiv leads to the formation of $[Sm(HMPA)_6]I_2$.[22] The combination of these findings is consistent with the equilibrium shown in Scheme 2.7. The mechanistic consequences of these findings are discussed in more detail in Chapter 3.

In one of the few examples of a chiral ligand-controlled asymmetric reaction mediated by SmI_2, Mikami reported the use of the chiral, bis-phosphine oxide, BINAPO (**4**), as a Lewis basic additive.[23] For example, reductive coupling of methyl acrylate and acetophenone using SmI_2 in THF with 2 equiv of BINAPO gave lactone **5** in 46% yield and 67% ee (Scheme 2.8).[23]

Although there are synthetic advantages to using HMPA in many SmI_2-mediated reactions, safety hazards make its use undesirable. Other Lewis basic cosolvents capable of acting as ligands for samarium have been utilised successfully in reactions with SmI_2; these include 1,3-dimethyl-3,4,5,6-tetrahydro-2(1*H*)-pyrimidinone (DMPU),[16,24,25] 1,1,3,3-tetramethylurea (TMU)[26] and nitrogen donor solvents.[27] DMPU has been shown to be more effective than HMPA as an additive in Julia–Lythgoe olefinations,[25] whereas the use of TMU with SmI_2 provided access to non-stabilised carbonyl ylides from iodomethyl silyl ethers.[26] In many of these reactions, a relatively large excess of additive is required. Although all of the aforementioned Lewis bases have found uses in specific reactions, none appear to provide the general utility of HMPA. Work by Flowers examined the impact of a number of basic cosolvents on the ease of oxidation of SmI_2. Although a number of cosolvents were found to increase dramatically the ease of oxidation of SmI_2 in THF, a large excess of the cosolvents was necessary compared with HMPA.[28] It was postulated that none of the cosolvents examined had as high an affinity for SmI_2 as HMPA. As a result, a large excess of alternative cosolvents was necessary to provide a fully ligated samarium complex. Crystallographic studies by Imamoto showed a

$$SmI_2(THF)_5 \overset{HMPA}{\rightleftharpoons} [Sm(THF)_2(HMPA)_4]I_2 \overset{HMPA}{\rightleftharpoons} [Sm(HMPA)_6]I_2$$

Scheme 2.7

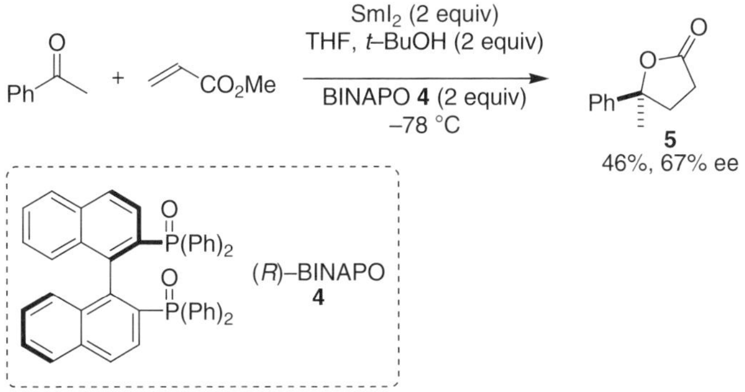

Scheme 2.8

large structural diversity for SmI_2–Lewis base complexes derived from TMU and DMPU.[29] Results from preliminary mechanistic studies on SmI_2-mediated reactions with other Lewis bases are consistent with coordination to Sm(II) being important for the production of a more reactive reductant in a manner similar to HMPA.

2.2.2 Proton Sources

Many reactions mediated by SmI_2 require the presence of a proton donor. The primary role of the proton donor is to quench alkoxides and carbanions produced as intermediates upon reduction or reductive coupling. The most commonly utilised proton donors are alcohols, glycols and water. It is now very clear, however, that proton donors can have a considerable impact on the efficiency of SmI_2-mediated reactions and their regiochemical and stereochemical outcome. Often, even a modest change in the proton donor or its concentration can have a profound impact on product distributions. Two important examples of this phenomenon are discussed below.

In 1999, Keck examined the reduction of β-hydroxy ketones using SmI_2 in THF, and screened a number of proton donors for use in the reaction.[30] Although low concentrations of water provided a good yield and reasonable diastereoselectivity in the conversion of **6** to **7**, an increase in water concentration resulted in a loss of diastereoselectivity. The use of methanol as a proton source provided excellent yields and diastereoselectivity over a range of concentrations whereas the use of *tert*-butanol led to the recovery of starting material (Scheme 2.9).

Another important example of the impact of a proton donor source on the reaction outcome was reported by Procter (Scheme 2.10).[31] γ,δ-Unsaturated ketones, such as **8**, undergo two very different, stereoselective cyclisation

OH O / Ph / **6** —[SmI_2, THF; proton source, 0 °C]→ OH OH / Ph / **7**

Proton source	Concentration of proton donor	Yield (%)	Ratio (*anti*:*syn*)
H_2O	2 equiv	96	83:17
H_2O	10 equiv	88	50:50
MeOH	2 equiv	95	98:2
MeOH	10 equiv	99	>99:1
t-BuOH	10 equiv	no rxn	–

Scheme 2.9

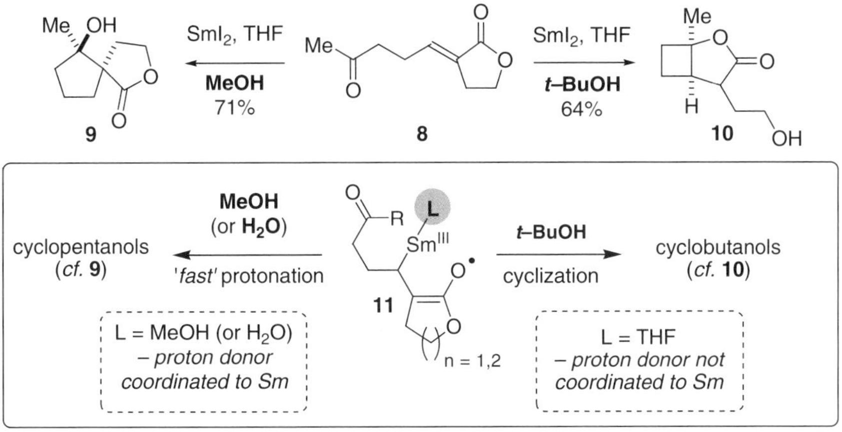

Scheme 2.10

reactions mediated by SmI_2 depending upon the alcohol cosolvent used in the reaction. Switching between a stereoselective five-membered ring-forming reaction, to give **9**, and a four-membered ring-forming reaction, to give **10**, was achieved simply by changing the alcohol cosolvent from MeOH to *t*-BuOH. Under either set of conditions, none of the alternative cyclisation product was isolated.[31] Although the mechanistic origin of this proton donor dependence is not yet clear, it is likely that it results from the differing rates of protonation of a radical-anion intermediate **11**. In MeOH, the radical-anion intermediate **11** is quenched rapidly by MeOH bound to the samarium centre, thus allowing the reaction to proceed to form cyclopentanols, such as **9**, whereas with *t*-BuOH a conventional protonation of the radical-anion intermediate **11** cannot compete with cyclisation to give cyclobutanol derivatives, such as **10** (Scheme 2.10).[31]

Oxygen-containing proton donors, including water, alcohols and glycols, coordinate to oxophilic SmI_2 and donate a proton to a reduced substrate through heterolytic cleavage of the O–H bond. If proton donor coordination to Sm(II) occurs along the reaction coordinate, the acidity of the proton donor will be enhanced. Curran first showed that water accelerated the rate of a number of SmI_2-mediated reductions and postulated that the reductant contained bound water.[24] A mechanistic study by Hoz showed that the reduction of activated alkenes by SmI_2 and proton donors occurred through two mechanistic pathways, one in which free SmI_2 reacted with substrate prior to protonation by an alcohol and another in which the Sm(II)–alcohol complex reacted with the substrate.[32] The reactant concentration, the proton source used and the mode of addition were all factors in determining the mechanistic pathway of reductions. Subsequent studies by Flowers on the impact of proton donors on the mechanism of SmI_2-mediated ketone reduction showed that the acidity of the proton donor and its affinity for SmI_2 played an important role in the mechanism of reduction.[33] In particular, water was found to have a higher affinity than alcohols for SmI_2, and generated a more powerful reductant.[34] The activation of SmI_2 by H_2O is best illustrated by Procter's recent findings that simple lactones can be reduced using SmI_2–H_2O.[35] The reduction of unfunctionalised aliphatic esters and lactones using SmI_2 had long been thought to be impossible.

The mechanistic studies discussed above led to the proposal that multidentate alcohols (such as glycols) capable of coordinating to SmI_2 should enhance the rate of reductions using the reagent. There are now a number of examples in the literature where ethylene glycol and related alcohols have been used to accelerate SmI_2-initated reductions[36] and deoxygenations.[37,38] Hilmersson has shown that coordinating alcohols enhanced the rate of ketone reduction substantially and the rate increase was proportional to the number of ethereal oxygens in the proton donor source.[39] Detailed studies by Flowers have shown that the mechanistic role of glycols is complex:[40] low concentrations of glycols enhance the reactivity of SmI_2 through displacement of solvent and iodide ligands from the inner sphere of Sm whereas high concentrations of glycols lead to coordinative saturation of Sm and provide a less reactive reductant.

Tetradentate chiral proton donors have been used for the asymmetric protonation of samarium enolates formed by the SmI_2 reduction of α-heteroatom-substituted carbonyl compounds. For example, Takeuchi examined the reduction of α-heterosubstituted cyclohexanone **12** using SmI_2 and the BINOL-derived chiral proton source **13**.[41] Ketone **14** was obtained in good yield and high enantiomeric excess (Scheme 2.11). Coordination of the proton source to samarium is key to the success of the transformation.[41]

It is clear that the ability of alcohols to protonate anionic reactive intermediates and their ability to coordinate to SmI_2 (thus increasing the effective molarity and acidity of the proton donor) are factors that should be considered when planning reaction sequences using the reductant.

In some cases, proton donors are used in concert with Lewis bases, such as HMPA. In most cases, these reactions provide predictable reaction outcomes with HMPA controlling the reactivity of SmI_2 and the proton donor having less

Scheme 2.11

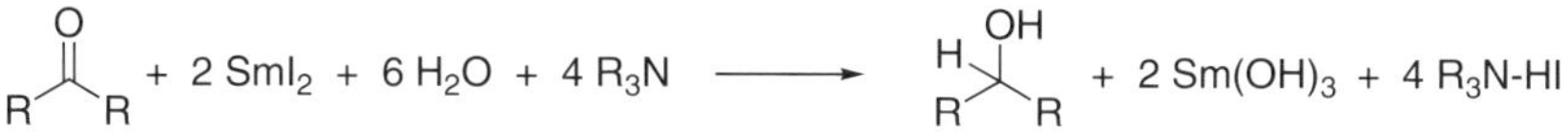

$$RC(O)R + 2\,SmI_2 + 6\,H_2O + 4\,R_3N \longrightarrow R_2CH(OH) + 2\,Sm(OH)_3 + 4\,R_3N\text{-}HI$$

Scheme 2.12

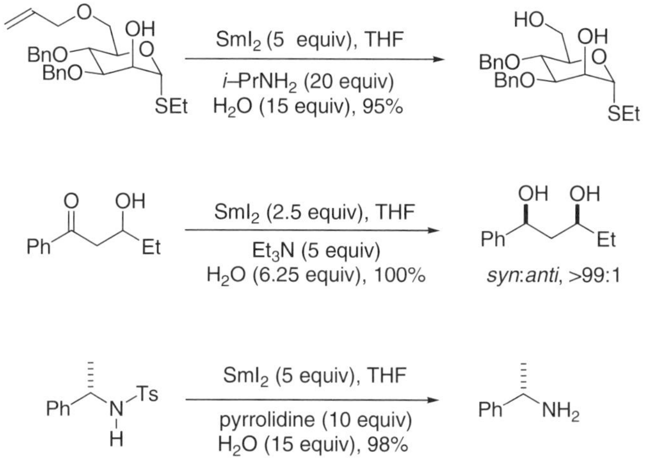

SmI$_2$ (5 equiv), THF; i-PrNH$_2$ (20 equiv), H$_2$O (15 equiv), 95%

SmI$_2$ (2.5 equiv), THF; Et$_3$N (5 equiv), H$_2$O (6.25 equiv), 100%; *syn*:*anti*, >99:1

SmI$_2$ (5 equiv), THF; pyrrolidine (10 equiv), H$_2$O (15 equiv), 98%

Scheme 2.13

influence. Fairly recently, Hilmersson showed that the combination of water and amines provides a unique method for enhancing the reactivity of SmI$_2$.[42] For example, the reduction of ketones proceeded efficiently when the appropriate stoichiometry of water and amine was used with SmI$_2$ (Scheme 2.12).[42]

This method not only provides an alternative to the use of HMPA and proton donors, but in many instances also results in very fast reactions. Another benefit of this approach is that the inorganic by-products of the reaction precipitate, making work-up straightforward. While the mechanism of the reaction has been shown to be complex, increased reactivity is certainly due to the high affinity of water for SmI$_2$. Formation of a SmI$_2$–H$_2$O complex reduces the pK_a of the bound water leading to deprotonation by amine. Subsequent precipitation of ammonium iodide drives the equilibrium of the reaction. This method has been used to reduce a wide range of substrates (Scheme 2.13).[43–45]

2.2.3 Inorganic Additives

Inorganic salts are another important additive used to enhance the rate and selectivity of SmI$_2$-mediated reactions. The use of inorganic additives can be traced to the seminal studies of Kagan during which he used catalytic amounts of ferric chloride to accelerate the coupling reactions of alkyl iodides and ketones.[1] The use of inorganic bases such as lithium amide, lithium methoxide and potassium hydroxide in conjunction with SmI_2 allowed the reduction of *aryl* esters, *aryl* carboxylic acids, *aryl* anhydrides and *aryl* amides to be carried out.[46] These substrates are not reduced by SmI_2 alone. Addition of lithium bromide and lithium chloride has been shown to accelerate the pinacol coupling of ketones.[47] In this case, it has been shown that chloride and bromide anions displace iodide from SmI_2 producing $SmCl_2$ and $SmBr_2$ *in situ* (see Chapter 7, Section 7.1).

Transition metal salts and complexes can also be an important component of reactions involving SmI_2. The most commonly utilised salts are based on Fe(III) or Ni(II) and Kagan has shown that NiI_2 was superior to other transition metal salts in many of these reactions.[48] As a result, NiI_2 has become the additive of choice in reactions requiring a transition metal-based catalyst. NiI_2 has been used in a range of important reactions including the conjugate addition of alkyl iodides to α,β-unsaturated esters, amides and lactones,[49,50] the coupling of acid chlorides and esters,[51] intramolecular cyclisations and Grob fragmentations,[52] and the coupling of alkyl halides with nitriles[53] (Scheme 2.14).

Although the addition of catalytic amounts of transition metal complexes and salts results in clear benefits in many SmI_2-mediated reactions, the mechanistic basis for their effect is unknown.[54] Examination of the experimental procedures utilising catalytic amounts of NiI_2 shows that, in most cases,

SmI_2 (3 equiv), t-BuOH (2 equiv), THF; 4 mol % NiI_2, 100%

1. SmI_2 (4 equiv), THF, 1 mol% NiI_2; 2. deoxygenated H_2O then $H_3O^{\oplus}$, 82% (98 : 2)

SmI_2 (2.5 equiv), THF, 2 mol % NiI_2; 0 °C to rt, 69%

SmI_2 (2.7 equiv), THF; 1 mol % NiI_2, 68%

Scheme 2.14

NiI_2 and SmI_2 are premixed before addition of the substrate. Evaluation of the E° value of the Ni(II)/Ni(0) redox couple clearly shows that SmI_2 is capable of readily reducing Ni(II) to Ni(0). As a result, it is possible that Ni(0) intermediates may be responsible for the unique chemistry initiated by the addition of catalytic amounts NiI_2 to SmI_2.

2.2.4 Miscellaneous Promoters

Alternative approaches for the acceleration of SmI_2-mediated reactions have been developed. For example, Ogawa found that the irradiation of a solution containing SmI_2 and a chloroalkane with light of wavelength 560–700 nm led to reduction to the corresponding alkane.[55] This approach was extended to other reactions, including sequential radical-anionic processes and carbonylations involving chloroalkanes and carbon monoxide. These initial findings were remarkable since alkyl and aryl chlorides are resistant to reduction by SmI_2 even in the presence of HMPA. Subsequent studies have shown that irradiation of SmI_2 in the visible range leads to electron transfer to substrates from an excited state of SmI_2.[56,57] Photoexcitation of SmI_2 has been used to promote a number of reactions, including ketone–nitrile cyclisations and the synthesis of (*Z*)-β,γ-unsaturated nitriles.[58,59] Microwave irradiation has also been shown to accelerate the reduction of a range of substrates by SmI_2, including ketones, imines, chloroalkanes and α,β-unsaturated esters.[60] Further studies are necessary to assess the generality of photo- and microwave irradiation for the promotion of SmI_2-mediated reactions.

2.2.5 Conclusions

It is now clear that additives and cosolvents can have a remarkable effect on SmI_2-mediated reactions. Effects can be relevantly straightforward, for example an improvement in reaction rate, yield or stereoselectivity, or can be dramatic, for example the enabling of an otherwise impossible reaction or a clean switch of reaction course. Although the importance of additives and cosolvents has been known for some time, only in recent years has significant progress been made in understanding the effects. This has largely been achieved through the application of a broad range of physical methods to the study of reaction mechanisms involving SmI_2. It is at this interface between organic synthesis and physical chemistry that some of the most exciting opportunities in the field can be found.

References

1. P. Girard, J.-L. Namy and H. B. Kagan, *J. Am. Chem. Soc.*, 1980, **102**, 2693.
2. G. A. Molander and C. Kenny, *J. Org. Chem.*, 1991, **56**, 1439.
3. T. Imamoto and M. Ono, *Chem. Lett.*, 1987, 501.

4. N. Akane, Y. Kanagawa, Y. Nishiyama and Y. Ishii, *Chem. Lett.*, 1992, 2431.
5. J. M. Concellón, H. Rodriguez-Solla, E. Bardales and M. Huerta, *Eur. J. Org. Chem.*, 2003, 1775.
6. J. A. Teprovich Jr., P. K. S. Antharjanam, E. Prasad, E. N. Pesciotta and R. A. Flowers II, *Eur. J. Inorg. Chem.*, 2008, 5015.
7. A. Dahlén and G. Hilmersson, *Eur. J. Inorg. Chem.*, 2004, 3020.
8. J.-L. Namy, M. Colomb and H. B. Kagan, *Tetrahedron Lett.*, 1994, **35**, 1723.
9. S. M. Ruder, *Tetrahedron Lett.*, 1992, **33**, 2621.
10. M. Kunishima, K. Hioki, T. Ohara and S. Tani, *J. Chem. Soc., Chem. Commun.*, 1992, 219.
11. B. Hamann-Gaudinet, J.-L. Namy and H. B. Kagan, *Tetrahedron Lett.*, 1997, **38**, 6585.
12. B. Hamann-Gaudinet, J.-L. Namy and H. B. Kagan, *J. Organomet. Chem.*, 1998, **567**, 39.
13. J. Inanaga, M. Ishikawa and M. Yamaguchi, *Chem. Lett.*, 1987, 1485.
14. K. Otsubo, K. Kawamura, J. Inanaga and M. Yamaguchi, *Chem. Lett.*, 1987, 1487.
15. A. Krief and A. M. Laval, *Chem Rev.*, 1999, **99**, 745.
16. G. A. Molander and J. A. McKie, *J. Org. Chem.*, 1992, **57**, 3132.
17. W. J. Evans, T. S. Gummersheimer and J. W. Ziller, *J. Am. Chem. Soc.*, 1995, **117**, 8999.
18. J. B. Shotwell, J. M. Sealy and R. A. Flowers II, *J. Org. Chem.*, 1999, **64**, 5251.
19. Z. Hou and Y. Wakatsuki, *J. Chem. Soc., Chem. Commun.*, 1994, 1205.
20. Z. Hou, Y. Zhang and Y. Wakatsuki, *Bull. Chem. Soc. Jpn.*, 1997, **70**, 149.
21. M. Shabangi and R. A. Flowers II, *Tetrahedron Lett.*, 1997, **38**, 1137.
22. R. J. Enemaerke, T. Hertz, T. Skrydstrup and K. Daasbjerg, *Chem. Eur. J.*, 2000, **6**, 3747.
23. K. Mikami and M. Yamaoka, *Tetrahedron Lett.*, 1998, **39**, 4501.
24. E. Hasegawa and D. P. Curran, *J. Org. Chem.*, 1993, **58**, 5008.
25. G. E. Keck, K. A. Savin and M. A. Weglarz, *J. Org. Chem.*, 1995, **60**, 3194.
26. M. Hojo, H. Aihara and A. Hosomi, *J. Am. Chem. Soc.*, 1996, **118**, 3533.
27. W. Cabri, I. Candiani, M. Colombo, L. Franzoi and A. Bedeschi, *Tetrahedron Lett.*, 1995, **36**, 949.
28. M. Shabangi, J. M. Sealy, J. R. Fuchs and R. A. Flowers II, *Tetrahedron Lett.*, 1998, **39**, 4429.
29. M. Nishiura, K. Katagiri and T. Imamoto, *Bull. Chem. Soc. Jpn.*, 2001, **74**, 1417.
30. G. E. Keck, C. A. Wager, T. Sell and T. T. Wager, *J. Org. Chem.*, 1999, **64**, 2172.
31. T. K. Hutton, K. W. Muir and D. J. Procter, *Org. Lett.*, 2003, **5**, 4811.
32. S. Hoz, A. Yacovan and I. Bilkis, *J. Am. Chem. Soc.*, 1996, **118**, 261.
33. P. Chopade, E. Prasad and R. A. Flowers II, *J. Am. Chem. Soc.*, 2004, **126**, 44.

34. E. Prasad and R. A. Flowers II, *J. Am. Chem. Soc.*, 2005, **127**, 18093.
35. L. A. Duffy, H. Matsubara and D. J. Procter, *J. Am. Chem. Soc.*, 2008, **130**, 1136.
36. A. Tarnopolsky and S. Hoz, *J. Am. Chem. Soc.*, 2007, **129**, 3402.
37. S. Hanessian and C. Girard, *Synlett*, 1994, 863.
38. R. J. Linderman, K. P. Cusakc and W. R. Kwochka, *Tetrahedron Lett.*, 1994, **35**, 1477.
39. A. Dahlén and G. Hilmersson, *Tetrahedron Lett.*, 2001, **42**, 5565.
40. J. A. Teprovich Jr, M. N. Balili, T. Pintauer and R. A. Flowers II, *Angew. Chem. Int. Ed.*, 2007, **46**, 8160.
41. Y. Nakamura, S. Takeuchi, Y. Ohgo, M. Yamaoka, A. Yoshida and K. Mikami, *Tetrahedron*, 1999, **55**, 4595.
42. A. Dahlén and G. Hilmersson, *Chem. Eur. J.*, 2003, **9**, 1123.
43. A. Dahlén, A. Sundgren, M. Lahmann, S. Oscarson and G. Hilmersson, *Org. Lett.*, 2003, **5**, 4085.
44. T. A. Davis, P. R. Chopade, G. Hilmersson and R. A. Flowers II, *Org. Lett.*, 2005, **7**, 119.
45. T. Ankner and G. Hilmersson, *Org. Lett.*, 2009, **11**, 503.
46. Y. Kamochi and T. Kudo, *Tetrahedron Lett.*, 1991, **32**, 3511.
47. J. R. Fuchs, M. L. Mitchell, M. Shabangi and R. A. Flowers II, *Tetrahedron Lett.*, 1997, **38**, 8157.
48. F. Machrouhi, B. Hamman, J.-L. Namy and H. B. Kagan, *Synlett*, 1996, 633.
49. G. A. Molander and C. R. Harris, *J. Org. Chem.*, 1997, **62**, 7418.
50. G. A. Molander and D. J. St. Jean Jr., *J. Org. Chem.*, 2002, **67**, 3861.
51. F. Machrouhi, J.-L. Namy and H. B. Kagan, *Tetrahedron Lett.*, 1997, **38**, 7183.
52. G. A. Molander, Y. Le Huérou and G. A. Brown, *J. Org. Chem.*, 2001, **66**, 4511.
53. H.-Y. Kang and S.-E. Song, *Tetrahedron Lett.*, 2000, **41**, 937.
54. H. B. Kagan, *J. Alloys Compd.*, 2006, **408–412**, 421.
55. A. Ogawa, Y. Sumino, T. Nanke, S. Ohya, N. Sonoda and T. Hirao, *J. Am. Chem. Soc.*, 1997, **119**, 2745.
56. W. G. Skene, J. C. Scaiano and F. L. Cozens, *J. Org. Chem.*, 1996, **61**, 7918.
57. E. Prasad, B. W. Knettle and R. A. Flowers II, *Chem. Eur. J.*, 2005, **11**, 3105.
58. G. A. Molander and C. N. Wolfe, *J. Org. Chem.*, 1998, **63**, 9031.
59. J. M. Concellón, H. Rodriguez-Solla, C. Simal, D. Santos and N. R. Paz, *Org. Lett.*, 2008, **10**, 4959.
60. A. Dahlén, E. Prasad, R. A. Flowers II and G. Hilmersson, *Chem. Eur. J.*, 2005, **11**, 3279.

CHAPTER 3

Mechanisms of SmI_2-mediated Reactions – the Basics

3.1 Introduction

The unique place held by SmI_2 in the arsenal of the synthetic chemist is a result of its versatility in mediating numerous important organic reactions, including reductions, reductive couplings and sequential reactions, all of which are described in subsequent chapters. Although numerous applications have been discovered for SmI_2, its scope in synthesis has not been exhausted and new applications for this mild and selective single-electron transfer reagent continue to be identified at a considerable rate. The aim of this chapter is to familiarise the reader with the basic mechanistic aspects of substrate reduction by SmI_2. This information is important for the rational design of new reactions mediated by the reagent. As most bond-forming reactions using SmI_2 generate radicals (and subsequently anions) through the reduction of organohalides, or ketyls, through reduction of carbonyl groups, the main emphasis will be on these processes as they represent the foundation for the reactions discussed in Chapters 4 and 5.

3.2 Radicals and Anions from Organohalides

The reduction of alkyl and aryl halides by SmI_2 provides access to radicals that can undergo a range of follow-up reactions, including dimerisation, reduction to an anion (organosamarium) or hydrogen atom abstraction from solvent as shown in Scheme 3.1.

Kagan's early mechanistic studies on the reduction of alkyl halides with SmI_2 suggested that organosamarium species were not intermediates in the reduction.[1] These studies were carried out by heating the substrate at reflux with SmI_2 and subsequently quenching the reaction mixture with D_2O. Alkane products were obtained with no deuterium incorporation, suggesting that alkyl radical intermediates were quenched by hydrogen atom abstraction from

Organic Synthesis Using Samarium Diiodide: A Practical Guide
By David J. Procter, Robert A. Flowers, II and Troels Skrydstrup

Published by the Royal Society of Chemistry, www.rsc.org

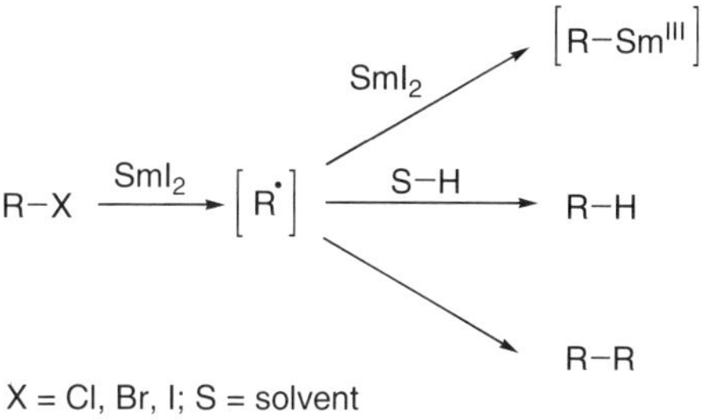

Scheme 3.1

THF.[1] However, when the reduction of 2-bromoadamantane was carried out at room temperature in the presence of D_2O using SmI_2–HMPA, the deuterated product was obtained in 80% yield. This result led to a comprehensive mechanistic study by Curran[2] that provided convincing evidence for the formation of organosamarium intermediates in the reduction of alkyl halides. Curran suggested that the lack of deuterium incorporation seen in Kagan's studies was a result of organosamarium decomposition at the elevated temperatures required.

The mechanism of the reduction of primary and secondary alkyl halides with SmI_2 in the presence of THF-HMPA is now known to proceed by dissociative electron transfer from SmI_2 to the alkyl halide, generating an alkyl radical **1**, followed by rapid reduction of the radical to an organosamarium species **2** that is then protonated to provide alkane **3** upon workup (Scheme 3.2).

The reduction of other carbon–halogen bonds is dependent on the substrate, solvent milieu and the presence of additives. Curran has shown that reduction of a tertiary alkyl radical, formed from a tertiary alkyl halide, can produce a tertiary organosamarium with limited stability at room temperature.[3] Most vinyl and aryl halides are reduced to radicals by SmI_2 and subsequently abstract a hydrogen atom from THF. This led Tani to prepare aryl organosamarium species by reduction of aryl halides using SmI_2–HMPA in benzene.[4] It is therefore clear that both aryl and alkyl radicals can be reduced by SmI_2 or SmI_2–HMPA, but hydrogen-atom abstraction from solvent and/or dimerisation can be a competing process.

Key to designing reactions using SmI_2 or SmI_2–HMPA is a fundamental understanding of the rates of reaction pathways and the seminal work of Curran is fundamentally important in this regard. Curran's detailed study of the reduction of alkyl halides provided rate constants for the reduction of primary, secondary and tertiary radicals with SmI_2–HMPA (Scheme 3.3).[2,5]

Hexenyl radicals were used as radical clocks for the indirect measurement of the rate of reduction of radicals to anions using SmI_2–HMPA. For example, reduction of primary iodide **4** using SmI_2–HMPA resulted in the isolation of coupled product **9** in 20% yield and cyclised-coupled product **7** in 80% yield. As the rate of cyclisation of the intermediate primary hexenyl radical **6** was known, a rate constant of $k = 10^6\ M^{-1}\ s^{-1}$ could be estimated for the reduction

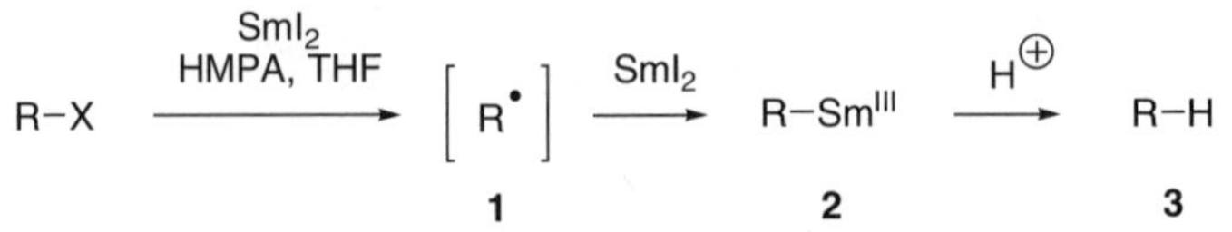

Scheme 3.2

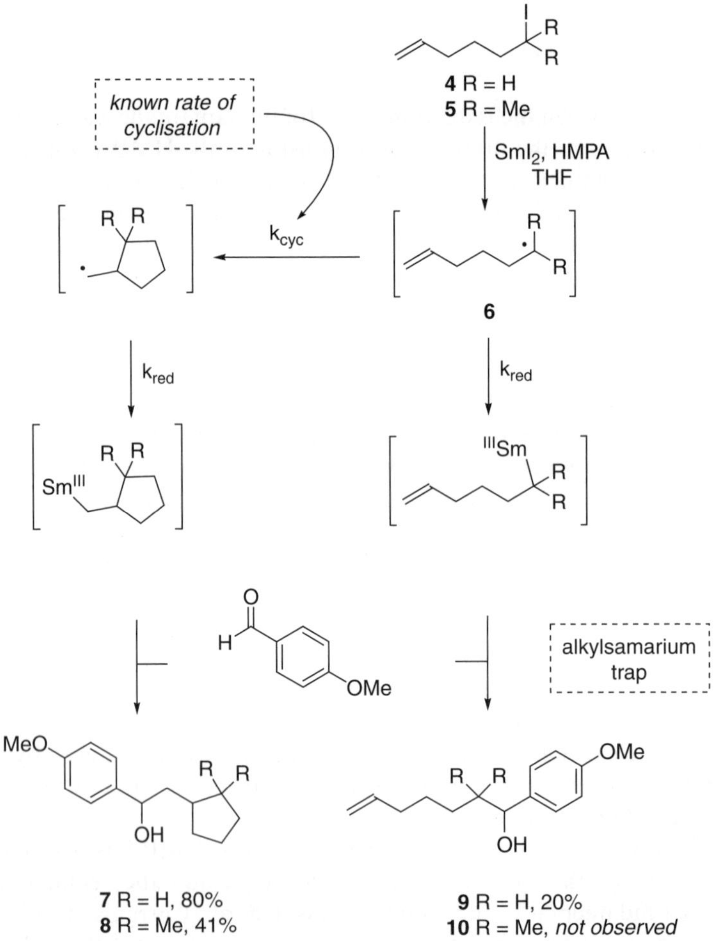

Scheme 3.3

of a primary alkyl radical to a primary alkylsamarium. Reduction of the tertiary alkyl iodide **5** using SmI_2–HMPA resulted in the isolation of cyclised–coupled product **8**, as the only product. The reduction of a tertiary alkyl radical to a tertiary alkylsamarium is therefore slower ($k \approx 10^4\,M^{-1}\,s^{-1}$) than the reduction of primary radicals under the same conditions. From these data, it can be inferred that the reduction of a secondary alkyl radical to a secondary alkylsamarium is intermediate in rate ($k \approx 10^5\,M^{-1}\,s^{-1}$).[2,5]

It is important to note that a sufficient quantity of HMPA is required for these reductions. The efficiency of reduction of **4** (shown in Scheme 3.3) was dependent on the amount of HMPA present. In the absence of HMPA, reduction of **4** was too slow for the rate experiments. Appropriate reaction times were obtained after addition of as little as 2 equiv of HMPA (based on [SmI_2]) and the ratio of **9** to **7** increased upon continued addition of HMPA (up to 5 equiv). These findings suggest that the rate of radical reduction can be fine-tuned by varying the amount of HMPA used. In practice, however, reactions that require the use of HMPA are rarely efficient if less than 4 equiv of HMPA are used (see Chapter 2, Section 2.2).

Curran's rate data for the reduction of alkyl radicals to the corresponding anions in the presence of SmI_2–HMPA is indispensable for the planning of radical and anionic transformations using the reagent: any proposed radical reaction using SmI_2 must be faster than the rate of reduction of the radical and, similarly, any proposed anionic process must take into account competition from fast radical reactions. Curran's approximate rate data are summarised in Scheme 3.4.[2]

The mechanistic significance of the reduction of alkyl radicals to organosamariums is well illustrated by the continued ambiguity surrounding the mechanism of the SmI_2-mediated Barbier reaction (see Chapter 5, Section 5.4). The SmI_2-mediated reductive couplings of an alkyl halide with a ketone or aldehyde are typically run using Barbier conditions, in which the substrates and SmI_2 are mixed together simultaneously in a reaction flask (Scheme 3.5).

Since both the alkyl halide and the carbonyl substrate can be reduced independently by SmI_2 or SmI_2–HMPA, mechanistic studies have been performed to determine whether the reaction occurs through the selective reduction of the alkyl halide to produce an organosamarium (*cf.* **2** in Scheme 3.2) or whether the reaction is the result of an alternative mechanism involving the

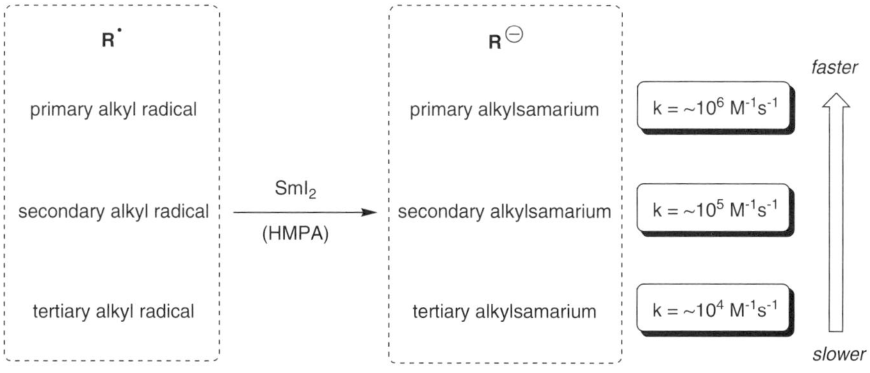

Scheme 3.4

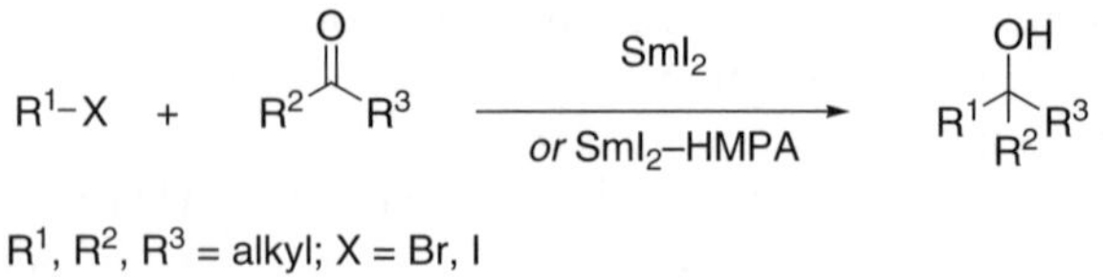

$R^1{-}X + R^2C(O)R^3 \xrightarrow[\text{or } SmI_2\text{–HMPA}]{SmI_2} R^1R^2R^3C{-}OH$

R^1, R^2, R^3 = alkyl; X = Br, I

Scheme 3.5

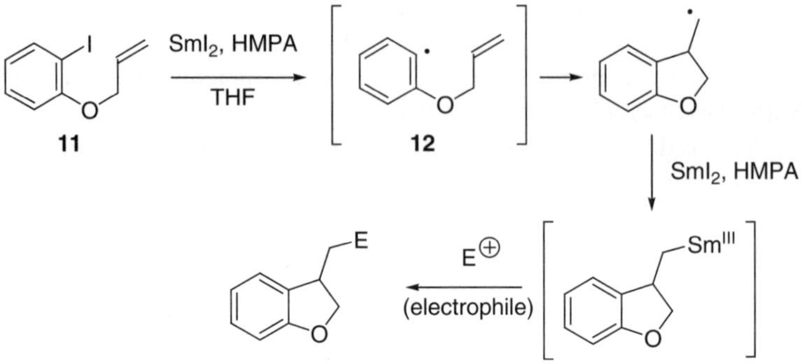

Scheme 3.6

initially formed alkyl radical (*cf.* **1** in Scheme 3.2).[2] Whereas the mechanism of the SmI_2-mediated Barbier reaction is likely to be substrate dependent, the reductive coupling of alkyl halides and ketones or aldehydes is now widely accepted to proceed through the intermediate formation of an organosamarium intermediate.[6,7] The synthetic applications of these reactions are described in detail in Chapter 5, Section 5.4.

As previously mentioned, fast radical processes, such as hydrogen atom abstraction from solvent or cyclisation, will compete with the reduction of a radical to an organosamarium. For example, Curran has shown that reduction of iodide **11** with SmI_2–HMPA generates aryl radical **12**, which cyclises to provide high yields of organosamarium intermediates that can be trapped with a range of electrophiles.[6] In this case, the radical cyclisation is efficient since it is several orders of magnitude faster than reduction of the intermediate aryl radical to an organosamarium intermediate (Scheme 3.6).

Although Curran's rate data for the reduction of radicals to organosamariums allow for an element of predictablity,[2] problems can arise when multifunctional substrates are involved. For example, in the attempted intramolecular Barbier reaction of alkyl iodide **13**, treatment with SmI_2 results in the formation of side product **15** in addition to the expected product cyclohexanol **14** (Scheme 3.7).[8] In this case, the β-keto amide motif in **13** is reduced at a rate competitive with alkyl iodide reduction, indicating that there are likely two mechanistic pathways through which the reaction proceeds: a thermodynamic pathway initiated by reduction of the R–I bond providing the

Scheme 3.7

cyclohexanol **14** and a kinetic pathway involving chelation of SmI_2 to the β-keto amide and reduction of the ketone carbonyl group to give **15**.[8]

The recognition of structural features in substrates that can potentially alter the course of reactions is important when planning reactions employing SmI_2.

3.3 Ketyl Radical Anions from Carbonyl Groups

The reduction of a carbonyl group by SmI_2 is the first step in a number of reactions of synthetic importance including reductions to alcohols, pinacol couplings and the coupling of carbonyls with alkenes. Numerous studies on the mechanism of ketone and aldehyde reduction by SmI_2 have been conducted and it is now recognised that additives, including HMPA and proton donors (*e.g.* alcohols and water), play an important role in the chemoselectivity and stereoselectivity of all reactions that are initiated by carbonyl reduction. Although the above effects can be considered as arising from 'reagent control', 'substrate control', for example when a neighbouring functional group in the substrate coordinates to SmI_2, can also be important. This section focuses on the basic mechanistic elements of carbonyl reduction as it relates to synthetically important reactions.

Studies on the reduction of carbonyl compounds using low-valent metallic reagents, including SmI_2, have led to an understanding of the mechanism involved. Single-electron transfer to the aldehyde or ketone generates a metal ketyl radical anion **16** that can form dimeric or polymeric ion pairs **17** as shown in Scheme 3.8.[9] Hoz carried out a detailed study on the reduction of substituted benzophenones and provided evidence that dissociation of Sm(III) from the dimer is an important mechanistic step leading to pinacol coupling.[9]

When considering the reduction shown in Scheme 3.8, it is important to keep a number of points in mind. The equilibrium in the initial reduction (step 1) lies to the side of starting material for dialkyl ketones and alkyl-substituted aldehydes. The presence of aryl groups stabilises the ketyl intermediate and, in the case of diaryl-substituted ketones (benzophenone derivatives), the equilibrium lies to the side of the ketyl and the reductions are fast. In a separate study on the reduction of benzophenones with SmI_2, Hoz suggested that the electrostatic interaction between the negatively charged radical anion and $Sm(III)^+$ is a major driving force behind the reduction.[10] Second, the addition of additives such as HMPA, proton donors or the presence of a proximal functional group

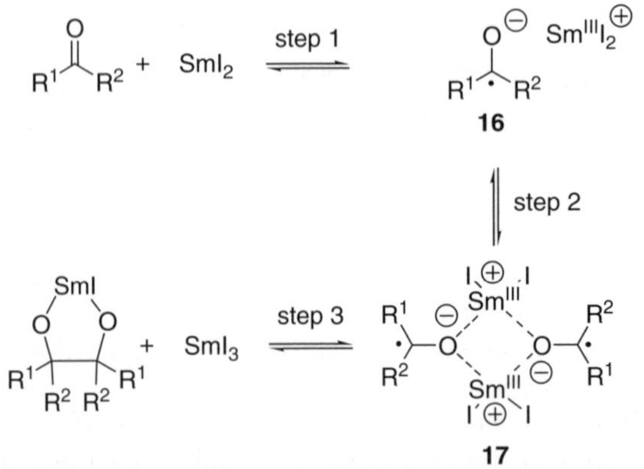

Scheme 3.8

capable of coordinating to SmI_2 also favours ketyl formation and accelerates the reduction of a carbonyl. The mechanistic impact of this second set of factors is discussed below.

3.3.1 The Role of Proton Donors in Carbonyl Reduction

Addition of proton donors (*e.g.* alcohols, glycols and water) to SmI_2 has a profound impact on the rate and mechanism of carbonyl reduction. Kinetic studies by Flowers showed that the rate of ketone reduction is directly related to the pK_a of the alcohol proton source used in the reaction and that the proton source must be sufficiently acidic to protonate the ketyl radical anion intermediate produced after the initial electron transfer.[11] Interestingly, when H_2O is used as a proton source, complexation of H_2O with SmI_2 results in a modified reagent and ketone reduction proceeds through a distinct pathway.[12] Hoz studied the role of 2,2,2-trifluoroethanol in the reduction of substituted benzophenones by SmI_2.[9] After electron transfer, ketyl **16** is in equilibrium with dimer **17** as shown in step 2 in Scheme 3.8. Spectral fitting analysis was used to analyse the rates of proton transfer from trifluoroethanol to the monomer (k_m) and dimer (k_d) (Scheme 3.9). Interestingly, Hoz found that protonation of the dimer was an order of magnitude faster than protonation of the monomeric ketyl.[9]

Careful consideration of Scheme 3.9 allows one to make reasonable suppositions about how to control further reactions of **16**. Once radical **16** is formed, the outcome of the reaction depends upon the relative stoichiometry of other components present in the reaction system. The use of a proton donor such as trifluoroethanol leads to rapid conversion of either dimeric or monomeric ketyl to intermediate **18**. Rapid production of **18** leads to dimerisation and formation of pinacol **19**. If protonation of the ketyl is slow, reduction of **18** by excess ketyl or SmI_2 followed by protonation leads to alcohol **21** through a traditional House-type mechanism (Scheme 3.10).[9]

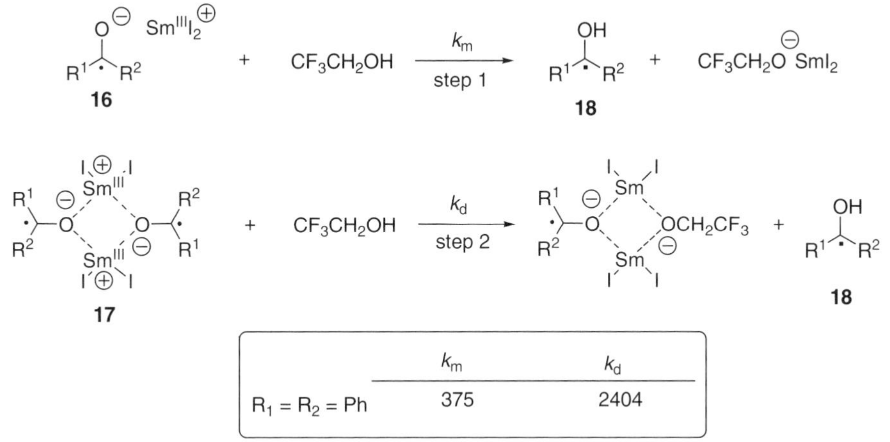

	k_m	k_d
$R_1 = R_2 = Ph$	375	2404

Scheme 3.9

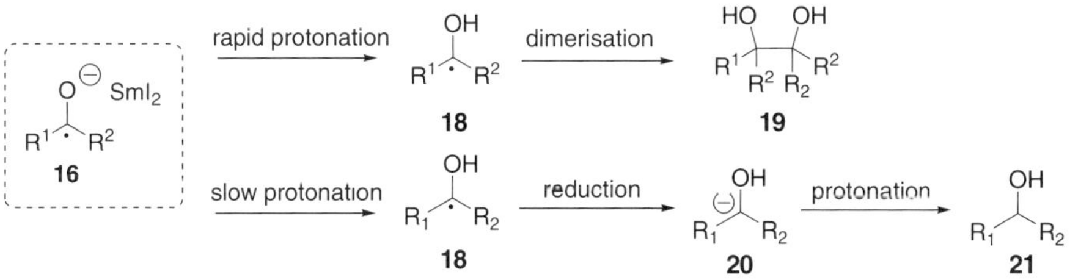

Scheme 3.10

Another factor that needs to be considered is the ability of proton donors to coordinate to SmI_2. While water has been shown to coordinate to SmI_2, other alcohols such as methanol also form complexes with SmI_2 and this coordination has an important impact on the mechanism of carbonyl reduction. The presence of a coordinated alcohol during reduction of a carbonyl by SmI_2 places the proton in close proximity to the developing negative charge on the ketyl radical anion, facilitating rapid proton transfer (Scheme 3.11). As a consequence, the reductions of less reactive carbonyl substrates are accelerated by the addition of an appropriate proton donor.

As proton donor coordination has an impact on the rate of carbonyl reduction, it is reasonable to expect that glycols should significantly accelerate the rate of carbonyl reduction. Seminal work by Hilmersson examined the correlation between the number of ethereal oxygens in a series of ethylene glycol-based proton donors and the rate of reduction of ketones.[13] His work showed that coordinating alcohols enhance the rate of ketone reduction substantially and that the rate increase is proportional to the number of ethereal oxygens in the proton donor source.[13] The mechanistic basis for this

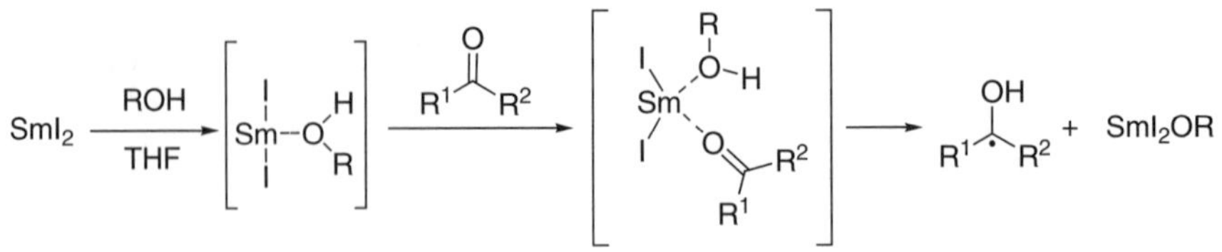

Scheme 3.11

acceleration is likely related to the proximity of the proton donor to the carbonyl being reduced, as described in Scheme 3.11. Interestingly, a large excess of glycol can, however, decrease the rates of some reactions. Flowers has shown that addition of glycol sufficient to fill the coordination sphere of SmI_2 actually decreases the rate of substrate reduction through the production of a sterically encumbered complex that is unable to coordinate to the ketone substrate.[14]

3.3.2 The Role of HMPA in Carbonyl Reduction

In Chapter 2, Section 2.2.1, we saw how the addition of HMPA to SmI_2 increases its reducing power and concomitantly produces a sterically congested reductant. As a consequence, the addition of HMPA to SmI_2 increases the rate of carbonyl reduction. Mechanistic work by Flowers has shown that the rate of reduction of dialkyl ketones increases by 1–2 orders of magnitude upon the addition of 4–10 equiv of HMPA.[15] The results of these studies are consistent with coordination of ketyl to Sm even in the presence of 10 equiv of HMPA.

Although reduction of dialkyl ketones and alkylaldehydes by SmI_2 occurs more quickly in the presence of HMPA through predictable pathways, care should be taken when carrying out reductions of arylaldehydes. Fang has shown that reduction of benzaldehyde derivatives provides 'head-to-tail' coupling products predominantly when HMPA is used as an additive with SmI_2 (Scheme 3.12).[16] Fang proposed that the presence of coordinated HMPA in ketyl intermediate **22** inhibits pinacol coupling. Instead, reduction provides an intermediate organosamarium species that reacts with another equivalent of benzaldehyde to furnish dianion **23** (R = H), before protonation gives 'head-to-tail'-coupled product **24** (R = H). Acetophenones also undergo similar dimerisation reactions upon reduction by SmI_2–HMPA whereas benzophenone derivatives do not.

Scheme 3.12 shows that although the addition of HMPA accelerates the electron transfer process in carbonyl reductions, it also plays an important role in post-electron transfer steps. Hoz studied the mechanistic role of HMPA in reactions involving the ketyl radical formed from 4,4′-dichlorobenzophenone.[17] It was found that increasing the amount of HMPA led to a decrease in the rate of coupling to give pinacol product. As shown in Scheme 3.8 (step 3), Sm(III) is required for bridging. Hoz proposed that increasing the amount of HMPA reduces the concentration of Sm(III) available for bridging through competitive coordination with metal, thus decreasing the rate of coupling.

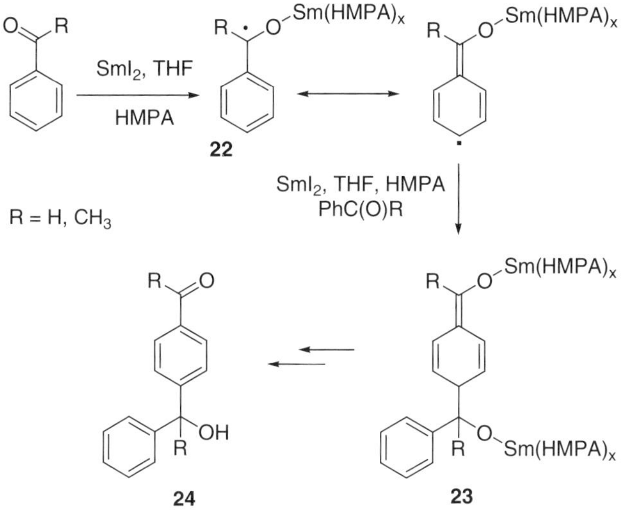

Scheme 3.12

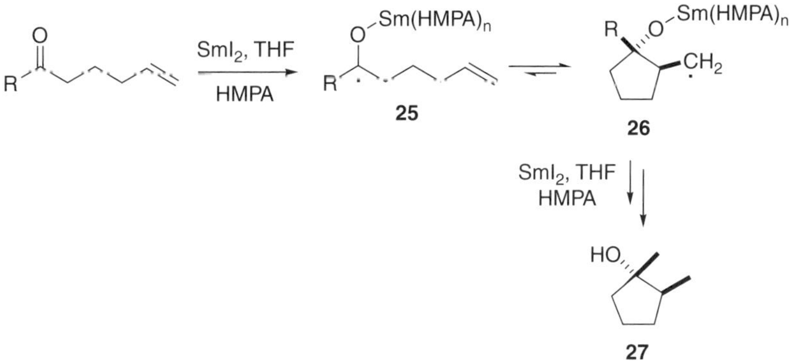

Scheme 3.13

HMPA has also been shown to play an important role in post-electron transfer steps in ketone–alkene coupling reactions. This carbon–carbon bond-forming reaction provides an important route to cyclic alcohols (see Chapter 5, Section 5.2) and the presence of HMPA is often important for obtaining high yields and diastereoselectivities in the transformation (see also Chapter 2, Section 2.2.1). Molander originally proposed that after initial reduction, Sm(III)-bound HMPA stabilises ketyl intermediate **25** through steric inhibition of competing reaction processes, such as hydrogen atom abstractions from solvent that lead to alcohol side products;[18] 5-*exo-trig* radical cyclisation of **25** then gives cyclopentanol **27** *via* radical **26** (Scheme 3.13).

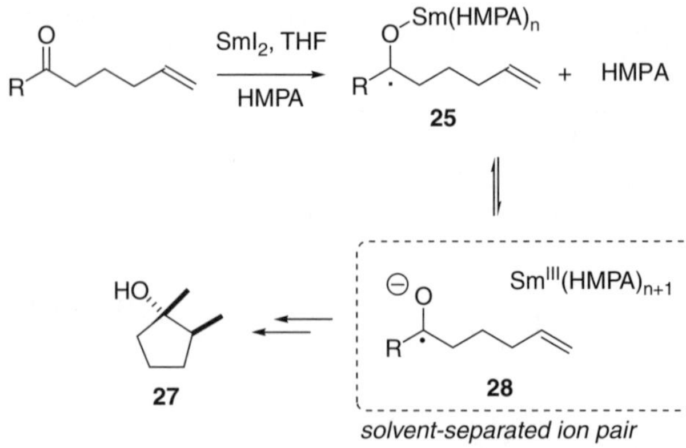

Scheme 3.14

A recent follow-up study by Flowers revealed that the role of HMPA is even more complex than initially proposed.[19] Using a series of rate and mechanistic studies, it was shown that HMPA plays a further mechanistic role after the initial electron transfer. Not only does coordination of HMPA to Sm(III) stabilise the intermediate ketyl **25**, but this coordination also inhibits cyclisation. Only upon liberation of the contact ion pair by another molecule of HMPA is solvent-separated ion pair **28** formed and the steric constraint to cyclisation removed (Scheme 3.14).[19]

3.3.3 The Role of Proximal, Lewis Basic Functional Groups in Carbonyl Reduction

From the examples discussed in the above sections, it is clear that additives (*e.g.* proton sources and HMPA) have a marked impact on the outcome of carbonyl reductions mediated by SmI_2, by subtly changing the nature of the reactive intermediates involved. The presence of a neighbouring functional group capable of chelation to SmI_2 can also have a significant impact on the rate and selectivity of carbonyl reduction. Appropriately placed functional groups provide a pathway for chelation (*versus* coordination), thus making the conversion of the carbonyl group to a ketyl more favourable (Scheme 3.15).

If chelation to a neigbouring group is possible, the use of HMPA is often not required for reduction of the carbonyl and reactions can be carried out under mild conditions. To examine the impact of chelation, Flowers studied the rate of reduction of 2-butanone, methyl acetoacetate and *N*,*N*-dimethylacetoacetamide by SmI_2.[20] Reduction of the β-keto ester or amide was several orders of magnitude faster than that of the unsubstituted ketone, which is consistent with chelation playing a major role. Further rate and mechanistic studies on the reduction of acetophenone and a series of 2′- and 4′-substituted acetophenone derivatives showed that both chelation and coordination provide highly

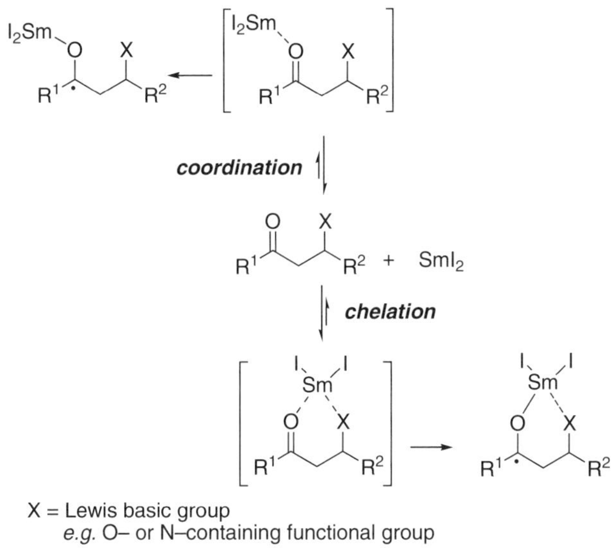

I2Sm
O
X
R1
R2
coordination
SmI2
chelation
Sm
I
X = Lewis basic group
e.g. O– or N–containing functional group

Scheme 3.15

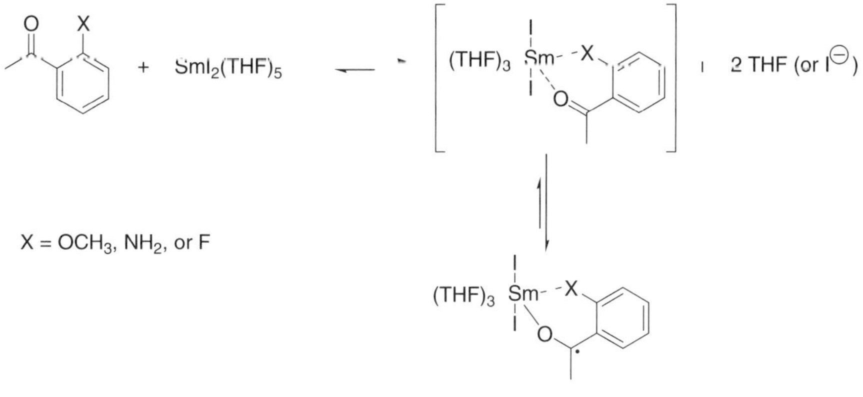

O
X
+ SmI2(THF)5
(THF)3
Sm
2 THF (or I⊖)
X = OCH3, NH2, or F

Scheme 3.16

ordered transition states for the reduction of a carbonyl to a ketyl, but the presence of a chelation pathway dramatically increased the rate of reduction. Interestingly, the data obtained in this study were consistent with solvent (or iodide) displacement providing the driving force for the rate enhancement (Scheme 3.16).[20]

In addition to accelerating the rate of carbonyl reduction, chelation can be used to control the diastereoselectivity of reductions and carbon–carbon bond-forming reactions through highly organised transition states. Keck showed that appropriately subsituted β-hydroxy ketones are stereoselectively reduced by

SmI_2 with methanol as a proton donor to give *anti*-diols under mild conditions.[21] Although proton donors have a marked impact on this reaction (Chapter 2, Section 2.2.2), Keck proposed that the key feature responsible for the observed selectivity is the intermediacy of the chelated intermediate **29** (Scheme 3.17). The organised structure provided through chelation leads to protonation of the most stable conformer containing a pseudoequatorial organosamarium. Further support for the importance of chelation is provided by the observation that the replacement of the β-hydroxyl group by sterically encumbered *tert*-butyldimethylsilyl and benzyl ethers prevents reduction whereas replacement with methyl, MOM, MTM and MEM ethers provides high levels of selectivity.[22]

Even though the model shown above is consistent with the stereochemical outcome, Flowers has shown that changing the reaction conditions or the substituents in the β-hydroxy ketone substrates can have an impact on the stereochemical outcome of the chelation-controlled reaction.[23,24] A more detailed discussion of the mechanism of this reduction can be found in Chapter 4, Section 4.3.

Elegant work by Molander has shown that chelation control is also important in the carbonyl–alkene cyclisations of β-keto esters and amides containing a pendant alkene.[25] Although the addition of HMPA is usually necessary for the intramolecular carbonyl–alkene couplings of ketones containing a pendant unactivated alkene, analogous intramolecular couplings involving β-keto esters and amides do not require the addition of HMPA and proceed under very mild conditions. Again, chelation facilitates reduction of the substrate and leads to an organised intermediate that reacts with high diastereoselectivity (Scheme 3.18).[25]

Although the above examples show the role played by substituents β to the ketone carbonyl, it is important to recognise that more distant, Lewis basic

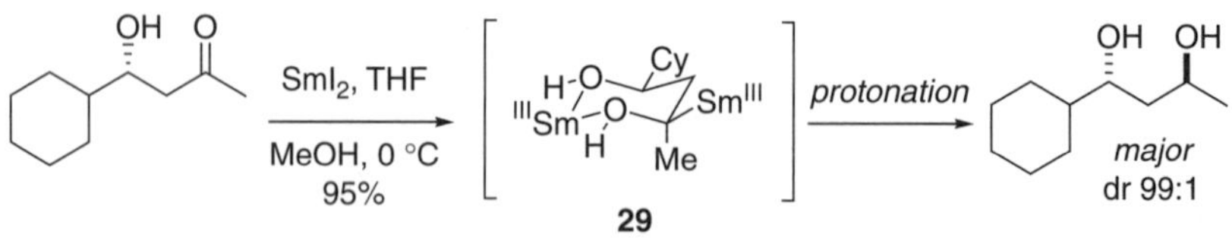

OH O

SmI_2, THF

MeOH, 0 °C

95%

H-O Cy ^{III}Sm-O Sm^{III} H Me

29

protonation

OH OH

major

dr 99:1

Scheme 3.17

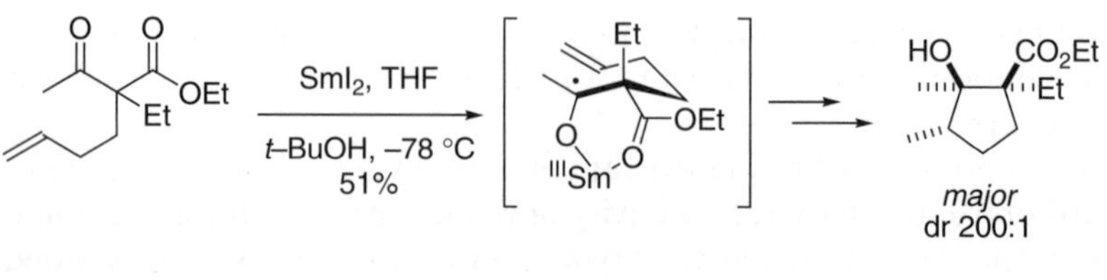

O O

OEt

Et

SmI_2, THF

t-BuOH, –78 °C

51%

Et

OEt

O O

^{III}Sm

HO CO_2Et

Et

major

dr 200:1

Scheme 3.18

functional groups can also coordinate to SmI_2, affecting reactivity and stereoselectivity. For example, eight- and nine-membered chelates have been invoked to rationalise the stereoselectivity observed in a number of reactions.[26,27] This facet of SmI_2-mediated chemistry should be kept in mind when employing the reagent in reactions involving highly functionalised substrates.

3.4 Mechanisms of Electron Transfer in Reactions Mediated by SmI_2

During the course of an electron transfer event, the electron can be transported from one substrate to another through one of two pathways: an outer-sphere (non-bonded) electron transfer or an inner-sphere (bonded) process. In an outer-sphere electron transfer, the coordination sphere of the donor and acceptor involved in the reaction remain intact and there is very little or, at most, a weak interaction between the components in the transition state. Conversely, in an inner-sphere electron transfer, the donor and acceptor interact through a ligand or atom and there is strong interaction between the components in the transition state.

Since Sm(II) is oxophilic, it has a high affinity for oxygen-containing functional groups. As already discussed, many reactions of carbonyl groups with SmI_2 proceed through a mechanistic pathway involving chelation or coordination of the carbonyl-containing functional groups to Sm(II) and as a result reactions most likely occur through an inner-sphere (bonded) pathway. In the case of the reduction of alkyl iodides and other halides, which have a lower affinity for SmI_2, the mode of electron transfer is less clear. Furthermore, the presence of sterically demanding ligands tends to promote outer-sphere electron transfer since these ligands inhibit interaction between electron donors and acceptors by coordinatively saturating the Sm(II) centres. As a result, the addition of HMPA can be used to alter the mechanism of electron transfer.

The first study designed to examine the mechanism of electron transfer was carried out by Daasbjerg and Skrydstrup.[28] In their ground-breaking work, the rate constant for the reduction of benzyl bromide by SmI_2 in THF was compared with the rate of reduction by a series of known outer-sphere donors having redox potentials close to that of SmI_2.[28] Although the reduction was not found to be purely outer-sphere, the electronic interaction in the transition state was estimated to be relatively small, and thus closer to an outer-sphere mechanism. Conversely, reduction of acetophenone by SmI_2 was shown to proceed with strong electronic coupling in the transition state, which is consistent with an inner-sphere process.[28]

Daasbjerg and Skrydstrup also showed that the addition of HMPA did not hinder the inner-sphere electron-donating ability of SmI_2 and that the inner-sphere electron transfer character of the transition state increases on proceeding from 1-iodobutane and benzyl bromide to benzyl chloride and acetophenone.[29] At first glance, this finding seems surprising as HMPA is a sterically demanding ligand that one might expect to limit coordination to the substrate. Daasbjerg and Skrydstrup's findings support a dynamic equilibrium for SmI_2–HMPA complexes where HMPA coordination to Sm(II) does produce a more sterically demanding reductant; however, coordination to HMPA also opens up coordination sites on Sm(II), through displacement of iodide, which the substrate can then access (see Chapter 2, Section 2.2.1).

Further work by Flowers examined the role of solvent polarity in the electron transfer process.[30] Inner-sphere electron transfer kinetics show a weak dependence on solvent polarity due to the considerable orbital overlap of the donor–acceptor pair in the transition state. In an outer-sphere process, changes in solvent polarity alter the energetics of electron transfer. The addition of excess HMPA, beyond that required to saturate SmI_2, resulted in a linear correlation to the rate of reduction for alkyl iodides, whereas no impact was observed on the rate of ketone reduction.[30] Thus the experiments showed a striking difference in the electron transfer mechanism for the substrate classes, which is consistent with the operation of an outer-sphere-type process for the reduction of alkyl iodides and an inner-sphere-type mechanism for the reduction of ketones.[30] These findings are consistent with the observations of Daasbjerg and Skrydstrup.[28,29]

From a practical perspective, the affinity of substrates for SmI_2 can be used as a rough guide for predicting the mechanism of electron transfer. The reduction of halides is not purely outer-sphere, the outer-sphere character decreases in the order R–I > R–Br > R–Cl. The reduction of carbonyl compounds having a high affinity for Sm(II) can be considered as proceeding through inner-sphere electron transfer.

References

1. H. B. Kagan, J. -L. Namy and P. Girard, *Tetrahedron*, 1981, **37**, 175.
2. D. P. Curran, T. L. Fevig, C. P. Jasperse and M. J. Totleben, *Synlett*, 1992, 943.
3. T. Nagashima, A. Rivkin and D. P. Curran, *Can. J. Chem.*, 2000, **78**, 791.
4. M. Kunishima, K. Hioki, K. Kono, T. Sakuma and S. Tani, *Chem. Pharm. Bull.*, 1994, **42**, 2190.
5. E. Hasegawa and D. P. Curran, *Tetrahedron Lett.*, 1993, **34**, 1717.
6. D. P. Curran and M. J. Totleben, *J. Am. Chem. Soc.*, 1992, **114**, 6050.
7. E. Prasad and R. A. Flowers II, *J. Am. Chem. Soc.*, 2002, **124**, 6895.
8. G. A. Molander, J. B. Etter and P. W. Zinke, *J. Am. Chem. Soc.*, 1987, **109**, 453.
9. H. Farran and S. Hoz, *J. Org. Chem.*, 2009, **74**, 2075.
10. H. Farran and S. Hoz, *Org. Lett.*, 2008, **10**, 4875.
11. P. R. Chopade, E. Prasad and R. A. Flowers II, *J. Am. Chem. Soc.*, 2004, **126**, 44.
12. E. Prasad and R. A. Flowers II, *J. Am. Chem. Soc.*, 2005, **127**, 18093.
13. A. Dahlén and G. Hilmersson, *Tetrahedron Lett.*, 2001, **42**, 5565.
14. J. A. Teprovich Jr., M. N. Balili, T. Pintauer and R. A. Flowers II, *Angew. Chem.*, 2007, **46**, 8160.

15. E. Prasad and R. A. Flowers II, *J. Am. Chem. Soc.*, 2002, **124**, 6895.
16. J.-S. Shiue, C.-C. Lin and J.-M. Fang, *Tetrahedron Lett.*, 1993, **34**, 335.
17. H. Farran and S. Hoz, *Org. Lett.*, 2008, **10**, 865.
18. G. A. Molander and J. A. McKie, *J. Org. Chem.*, 1992, **57**, 3132.
19. D. V. Sadasivam, P. K. S. Antharjanam, E. Prasad and R. A. Flowers II, *J. Am. Chem Soc.*, 2008, **130**, 7228.
20. E. Prasad and R. A. Flowers II, *J. Am. Chem. Soc.*, 2002, **124**, 6357.
21. G. E. Keck, C. A. Wager, T. Sell and T. T. Wager, *J. Org. Chem.*, 1999, **64**, 2172.
22. G. E. Keck and C. A. Wager, *Org. Lett.*, 2000, **2**, 2307.
23. T. A. Davis, P. R. Chopade, G. Hilmersson and R. A. Flowers II, *Org. Lett.*, 2005, **7**, 119.
24. P. Chopade, T. A. Davis, E. Prasad and R. A. Flowers II, *Org. Lett.*, 2004, **6**, 2685.
25. G. A. Molander and C. Kenny, *J. Am. Chem. Soc.*, 1989, **111**, 8236.
26. G. A. Molander and J. A. McKie, *J. Am. Chem. Soc.*, 1993, **115**, 5821.
27. D. J. Edmonds, K. W. Muir and D. J. Procter, *J. Org. Chem.*, 2003, **68**, 3190.
28. R. J. Enemærke, K. Daasbjerg and T. Skrydstrup, *Chem. Commun.*, 1999, 343.
29. R. J. Enemærke, T. Hertz, K. Daasbjerg and T. Skrydstrup, *Chem. Eur. J.*, 2000, **6**, 3747.
30. E. Prasad and R. A. Flowers II, *J. Am. Chem. Soc.*, 2002, **124**, 6895.

CHAPTER 4

Functional Group Transformations Using SmI_2

4.1 Introduction

Functional group interconversions constitute the foundation upon which the discipline of organic chemistry is built. Reductants are one of the most fundamental classes of reagent used to carry out functional group transformations and SmI_2 is one of the most versatile members of this class. The SmI_2-mediated reduction of a number of functional groups, often in the presence of other groups, underpins the reagent's ability to mediate carbon–carbon bond-forming reactions, where reactive intermediates are formed from functional groups and trapped. The Lewis acidic nature of SmI_2 means that coordination to functional groups facilitates electron transfer. In addition, the Sm(III) salts that are produced as by-products can enhance the leaving ability of those groups to which they can complex, therefore facilitating their loss. As discussed in Chapter 2, Section 2.2, additives and cosolvents can be used to fine-tune the reactivity of the reagent according to the functional group that requires reduction. Finally, although they are powerful reducing systems, Sm(II) species are stable in alcohols, water and even acidic and basic aqueous solutions. This allows reactive intermediates to be quenched rapidly by additives before alternative, undesired pathways are accessed. The main functional group interconversions that can be carried out using the lanthanide reagent are summarised in Figure 4.1 and are discussed in detail in the following Sections.

4.2 The Reduction of Alkyl Halides with SmI_2

SmI_2 reduces alkyl halides with the ease of reduction following the expected order: iodides > bromides > chlorides. In Kagan's seminal publications on the use of SmI_2 in synthesis, he discussed the reagent's ability to reduce alkyl halides.[1] In these early studies, the reactions were carried out in THF at reflux and, under these conditions, the reduction of primary alkyl bromides and iodides to the corresponding alkanes proceeded with high yields.

Organic Synthesis Using Samarium Diiodide: A Practical Guide
By David J. Procter, Robert A. Flowers, II and Troels Skrydstrup

Published by the Royal Society of Chemistry, www.rsc.org

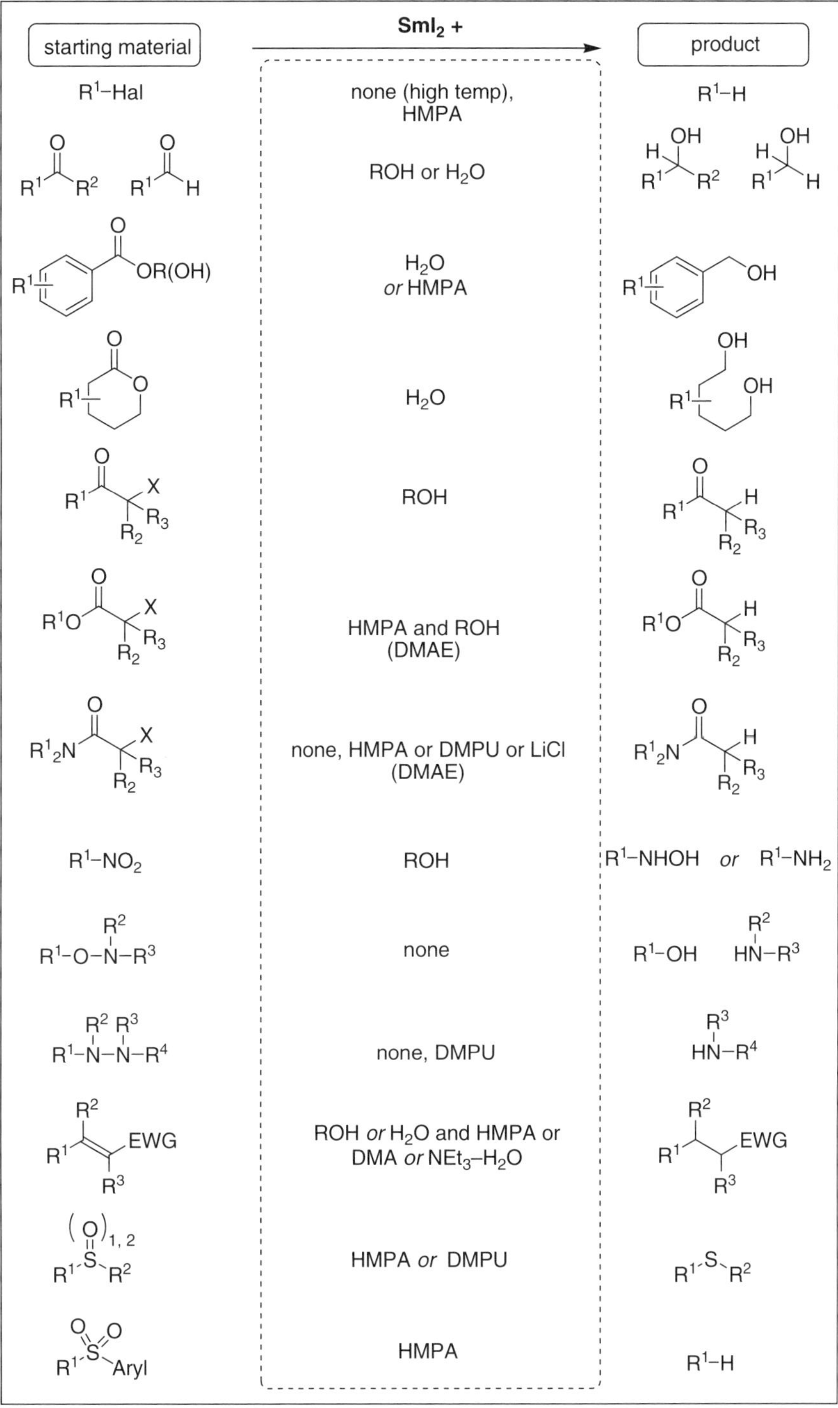

Figure 4.1 The main functional group interconversions that can be carried out using SmI_2.

$n\text{-}C_{10}H_{21}\text{–}Br \xrightarrow[\text{48 h, 95\%}]{SmI_2,\ THF} n\text{-}C_{10}H_{21}\text{–}H$

$n\text{-}C_{10}H_{21}\text{–}Cl \xrightarrow{SmI_2,\ THF}$ no reaction

$n\text{-}C_{10}H_{21}\text{–}Br \xrightarrow[>95\%]{SmI_2,\ \mathbf{HMPA},\ THF,\ 10\ min} n\text{-}C_{10}H_{21}\text{–}H$

$n\text{-}C_{10}H_{21}\text{–}Cl \xrightarrow[\text{8 h, >95\%}]{SmI_2,\ \mathbf{HMPA},\ THF,\ 60\ °C} n\text{-}C_{10}H_{21}\text{–}H$

Scheme 4.1

Inanaga found that the more powerful reductant obtained by the addition of HMPA to SmI_2 allowed the reduction of alkyl halides to be carried out under milder conditions (Scheme 4.1).[2] In the absence of HMPA, at room temperature, alkyl chlorides are not reduced and alkyl bromides react only slowly. On addition of the additive to SmI_2, alkyl bromides are reduced in minutes at room temperature and alkyl chlorides are converted to the corresponding alkanes upon heating. A detailed discussion of the mechanism of alkyl halide reduction can be found in Chapter 3, Section 3.2.

Representative procedure.[1] To a solution of SmI_2 in THF (2 equiv) under nitrogen was added the alkyl bromide (1 equiv) in THF. The reaction mixture was heated at reflux until the solution decolourised. The solution was quenched with 1 M HCl and the aqueous layer was extracted with Et_2O. The organic layers were combined and washed with water, sodium thiosulfate and brine before being dried with $MgSO_4$. The solvent was removed *in vacuo* and the products analysed by GLC.

Representative procedure.[2] To a 0.1 M solution of SmI_2 solution in THF (2.5 equiv) under nitrogen was added HMPA (5% of SmI_2 THF solution) and 2-propanol (1 equiv). To this was added the alkyl bromide (1 equiv) and the reaction mixture was stirred at room temperature. The products were analysed by GLC.

The ability of SmI_2 to reduce alkyl halides has been exploited in a number of carbon–carbon bond-forming reactions. Radicals generated from the reduction of alkyl halides can be trapped by alkenes in cyclisation reactions to form carbocyclic and heterocyclic rings (see Chapter 5, Section 5.3), and the alkylsamarium intermediates can be used in intermolecular and intramolecular Barbier and Grignard reactions (see Chapter 5, Section 5.4). The reduction of α-halocarbonyl compounds with SmI_2 gives rise to Sm(III) enolates that can be exploited in Reformatsky reactions (Chapter 5, Section 5.5) and are discussed in Section 4.5.

4.2.1 Generality and Scope of Alkyl Halide Reduction

In the presence of HMPA, SmI_2 is a useful reagent for the reduction of alkyl halides. Primary, secondary and tertiary alkyl halides and alkenyl

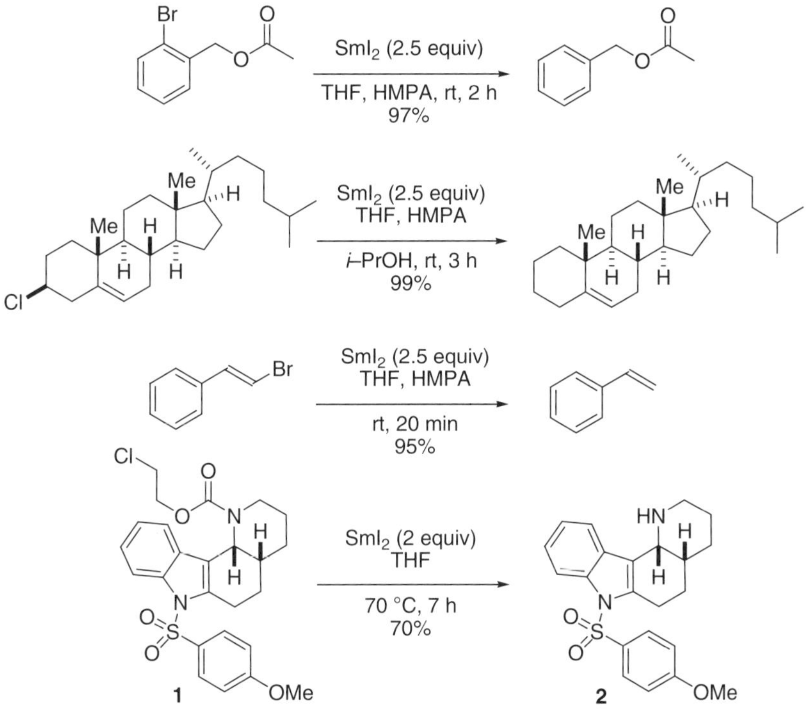

Scheme 4.2

and aryl halides are reduced rapidly under these conditions.[2–5] The reductions can be carried out in the presence of other functional groups.[2] Examples of selective alkyl, aryl and alkenyl halide reductions are shown in Scheme 4.2.

In the final example in Scheme 4.2, Magnus used SmI_2 to remove a carbamate protecting group from nitrogen in **1** by reduction of the alkyl chloride at elevated temperature.[6] A screen of alternative reducing agents for the deprotection showed SmI_2 to be by far the most effective.

4.2.2 β-Elimination Reactions by SmI_2 Reduction of Alkyl Halides

The reduction of alkyl halides with SmI_2 has been used to trigger β-elimination and the formation of alkenes.[7] For example, Concellón reported the reduction of 2-acetoxy-1,1-dihaloalkanes[8] and 2-acetoxy-1-chlorosilanes[9] to give vinyl halides and vinylsilanes, respectively (Scheme 4.3). In the reduction of dihaloalkanes bearing two different halides, only the more reactive halide ($I > Br > Cl$) is reduced.[7]

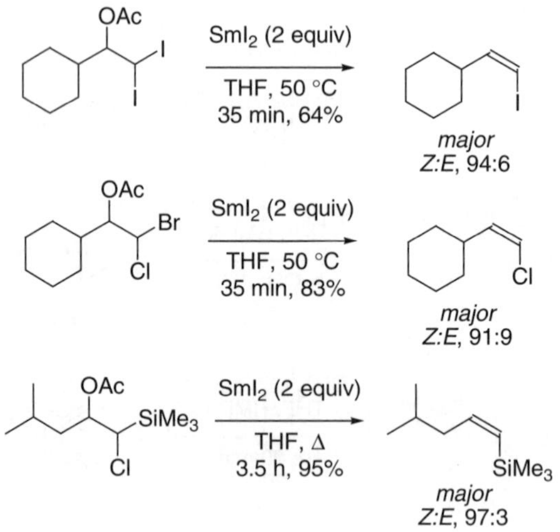

Scheme 4.3

4.3 The Reduction of Ketones and Aldehydes with SmI_2

SmI_2 reduces aldehydes and ketones to the corresponding alcohols. As there are numerous hydride reducing agents for the reduction of aldehydes and ketones, this functional group transformation has not been used widely. In some cases, however, SmI_2 can display useful reactivity, stereoselectivity and chemoselectivity that more conventional reagents cannot achieve.

4.3.1 The Mechanism of Carbonyl Reduction with SmI_2

Studies on the reduction of carbonyl compounds using low-valent metallic reagents, including SmI_2, have led to an understanding of the mechanism involved. Single electron transfer to the aldehyde or ketone generates a metal ketyl radical anion **3** that can form dimeric or polymeric ion pairs **4** (Scheme 4.4).[10]

In the presence of a proton source, the O–M bond of the metal ketyl radical anion is protonated to form a carbinol radical **5**. Further reduction then forms a hydroxyalkyl carbanion **6** and protonation gives an alcohol product (Scheme 4.5). Competing processes include dimerisation of the metal ketyl radical anion to give pinacol products or disproportionation to give an enolate and an alkoxide (Scheme 4.5).[10]

In a detailed kinetic study, Flowers showed that the rate of ketone reduction is directly related to the pK_a of the alcohol proton source used in the reaction and that the proton source must be sufficiently acidic to protonate the ketyl radical anion intermediate 3.[11] Interestingly, when H_2O is used as a proton

Scheme 4.4

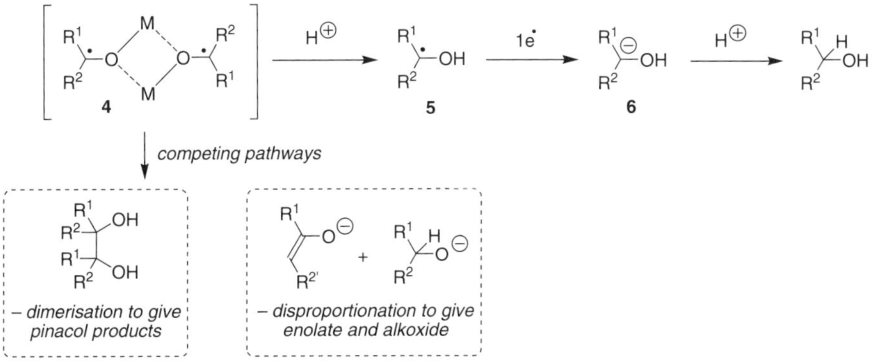

Scheme 4.5

source, complexation of H_2O to SmI_2 results in a modified reagent and ketone reduction proceeds through a distinct pathway.[11] Further discussion of the mechanistic details of carbonyl reduction and the impact of additives can be found in Chapter 3, Section 3.3.

The ketyl radical anion intermediates can be exploited in carbon–carbon bond-forming reactions. Intermolecular and intramolecular pinacol couplings between the carbonyl groups of ketones and aldehydes are well known (Chapter 5, Section 5.1), as are intermolecular and intramolecular carbonyl-alkene couplings (Chapter 5, Section 5.2).

4.3.2 Generality and Scope of Carbonyl Reduction with SmI_2

Ketone reduction using SmI_2 can be highly diastereoselective. For example, in Corey's synthesis of the diterpenoid natural product atractyligenin, SmI_2-mediated reduction of ketone **7** proceeded with complete diastereocontrol to give secondary alcohol **8** in 90% yield (Scheme 4.6).[12]

Keck reported that the reduction of β-hydroxy ketones with SmI_2 is an excellent method for the preparation of 1,3-*anti*-diols.[13] For example, reduction of **9** with SmI_2 in THF with MeOH gave 1,3-*anti*-diol **10** in 95% yield and 98% de. The reaction proceeds by initial coordination of Sm(II) to the free hydroxyl, followed by intramolecular electron transfer to the ketone carbonyl to form ketyl radical anion **11**. A second electron transfer gives alkylsamarium

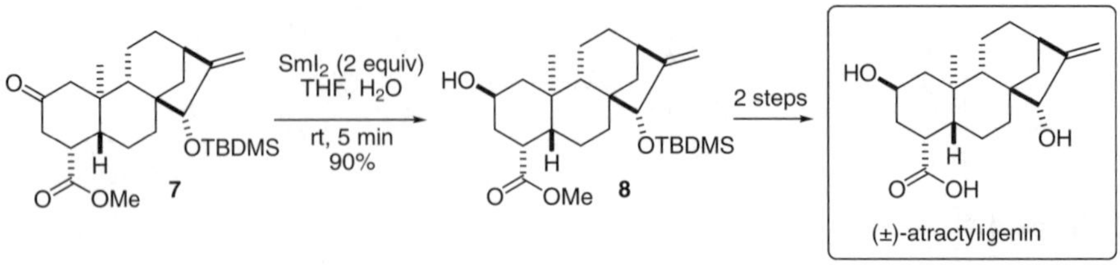

Scheme 4.6

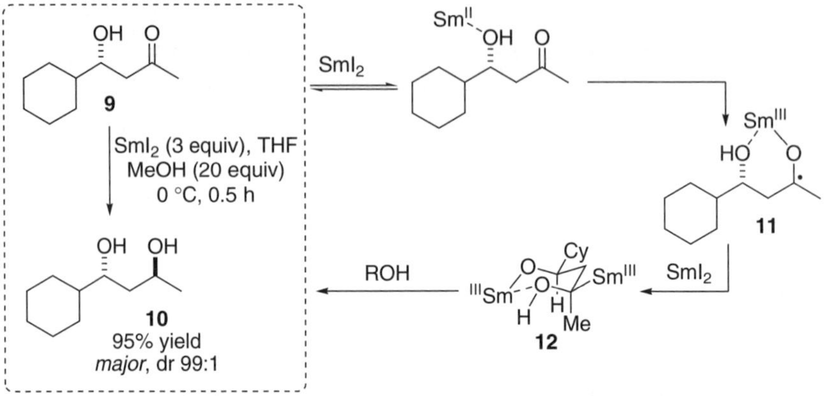

Scheme 4.7

12, in which the C–Sm bond adopts a pseudoequatorial position due to the steric demand of samarium and its associated ligands, before being protonated by MeOH to give **10** (Scheme 4.7).[13]

Interestingly, reduction of *tert*-butyl ketone **13** under the same conditions gave 1,3-diol **14** as a 1:1 mixture of diastereoisomers. Keck argued that samarium and its associated ligands are similar in size to the *tert*-butyl group and there is no preference for one organosamarium configuration over the other (Scheme 4.8).[13]

Keck showed that protection of the β-hydroxyl group in β-hydroxy ketones as *tert*-butyldimethylsilyl and benzyl ethers shut off reduction and starting material was recovered. This underlines the importance of pre-coordination of Sm(II) to the free hydroxyl in the directed reduction.[13] The *anti*-selective reduction does, however, tolerate methyl, MOM, MTM and MEM ethers, presumably as these small ether groups still allow coordination to Sm(II).[14] Keck pointed out that care should be taken when working with different systems as selectivities can be very substrate dependent.[13] Flowers has also shown that the diastereoselectivity of the reduction can be altered by changing the reaction conditions through the use of other solvents or reaction protocols.[15,16]

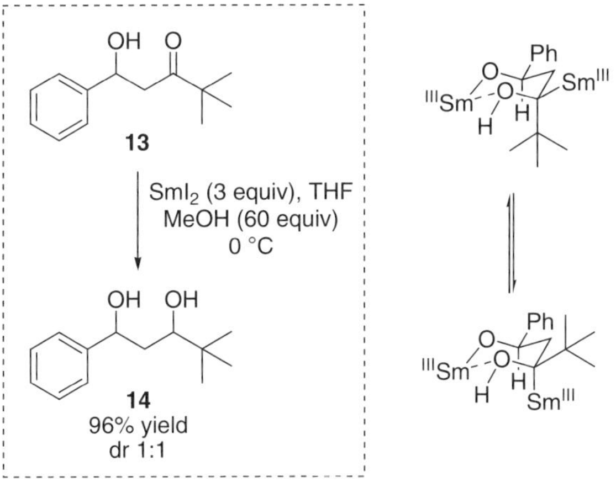

Scheme 4.8

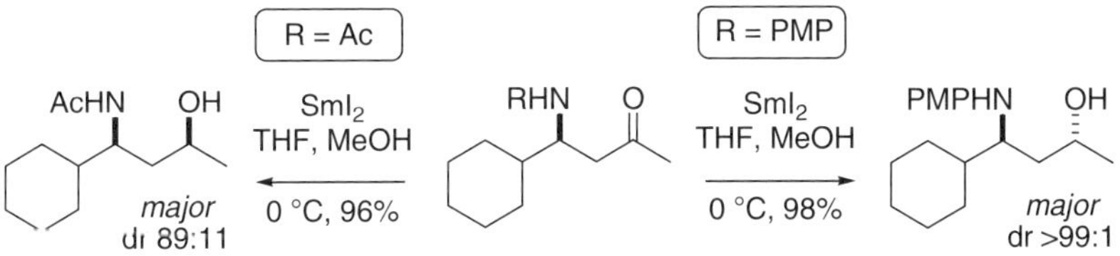

Scheme 4.9

Representative procedure.[13] A stirred solution of the β-hydroxy ketone (1 equiv) and MeOH (20 equiv) in THF under argon was cooled to 0 °C and 0.1 M SmI_2 solution in THF (3 equiv) was added dropwise. The reaction mixture was stirred until the blue solution decolourised. The reaction mixture was diluted with Et_2O and 1 M HCl and stirred for 10 min. The organic layer was separated and the aqueous layer extracted further with Et_2O. The combined organic extracts were washed with saturated aqueous $Na_2S_2O_3$, dried and the solvent removed *in vacuo*. The crude product was purified by column chromatography (EtOAc–hexane eluent).

Keck also reported the diastereoselective reduction of β-amino ketones using SmI_2 in THF with MeOH as the proton source.[17] In this case, reduction of *N*-acyl derivatives gave 1,3-*syn*-amino alcohols, whereas the reduction of *N*-aryl derivatives afforded 1,3-*anti*-amino alcohols (Scheme 4.9).[17]

Keck recently applied the diastereoselective reduction of β-hydroxy ketones in an approach to the anti-tumour natural products epothilone A and B. SmI_2-mediated reduction of β-hydroxy ketone **15** in THF–MeOH gave *anti*-1,3-diol **16** in 94% yield and as a 96:4 mixture of diastereoisomers.[18] Attempts to carry out the reduction using common hydride reducing agents gave **16** with poor levels of diastereoselectivity (Scheme 4.10).[18]

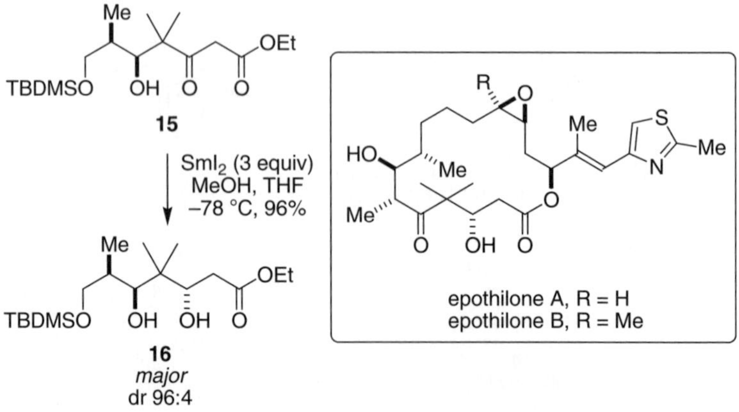

Scheme 4.10

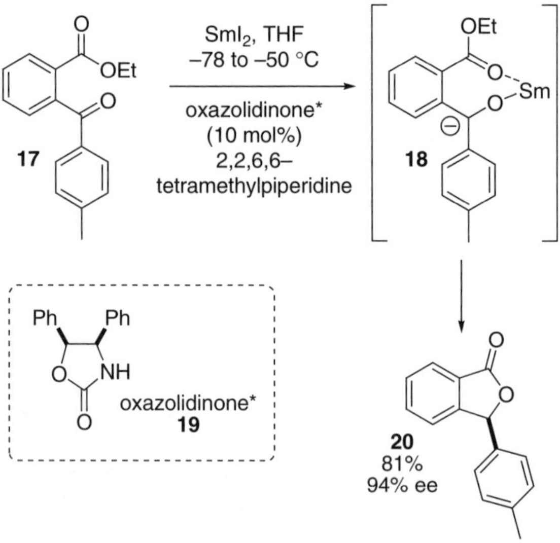

Scheme 4.11

Finally, in 2006, Xu and Lin reported an asymmetric reduction of 2-acylarylcarboxylates using SmI_2 and a catalytic amount of a chiral proton source.[19] For example, reduction of **17** gave chelated anion **18** that was protonated by enantiomerically pure oxazolidinone **19**. A stoichiometric, achiral proton source, 2,2,6,6-tetramethylpiperidine, then regenerated the chiral proton source. Lactone **20** was obtained in excellent yield and high enantiomeric excess (Scheme 4.11).[19]

4.3.3 Fragmentation Reactions Triggered by Carbonyl Reduction with SmI_2

The SmI_2-mediated reduction of ketones to give ketyl radical anions can result in the cleavage of a carbon–carbon bond α to the carbonyl group. This is a particularly efficient process when the carbon–carbon bond in question is part of a strained ring system. In fact, a cyclopropane group attached to a carbonyl group is often used to probe the mechanism of SmI_2-mediated reactions – if a ketyl radical anion is formed during the reaction, fragmentation of the three-membered ring occurs rapidly.[20,21] This fragmentation process is mechanistically related to the reduction of α-heteroatom-substituted carbonyl compounds using SmI_2 and proceeds *via* Sm(III) enolate intermediates (see Section 4.5 and Chapter 5, Section 5.5).

Kuwajima employed the fragmentation of cyclopropyl ketone **21** to introduce the C19-methyl group in an approach to the anti-cancer natural product paclitaxel (Taxol): treatment of **21** with SmI_2 gave enol **22** in quantitative yield (Scheme 4.12).[22]

Motherwell reported that the Sm(III) enolates or radicals resulting from cyclopropane fragmentation can be exploited in carbon–carbon bond-forming reactions.[23] For example, treatment of cyclopropyl ketone **23** with SmI_2 gave bicyclic ketone **24** from interception of an intermediate radical, while exposure of cyclopropyl ketone **25** to the lanthanide reagent gave **26** after quenching of the Sm(III) enolate intermediate with allyl bromide (Scheme 4.13).[23]

The fragmentation of cyclobutyl ketones has also been exploited in natural product synthesis. For example, Sorensen utilised the SmI_2-mediated fragmentation of cyclobutyl ketone **27** in a synthesis of guanacastepene A and E:

OH OH
TBSO
O O O
21 Ph
SmI_2
HMPA, MeOH
THF, 100%
OH OH
TBSO
O O OSm^{III}
Ph
OH OH
TBSO
O O OH
22 Ph

Scheme 4.12

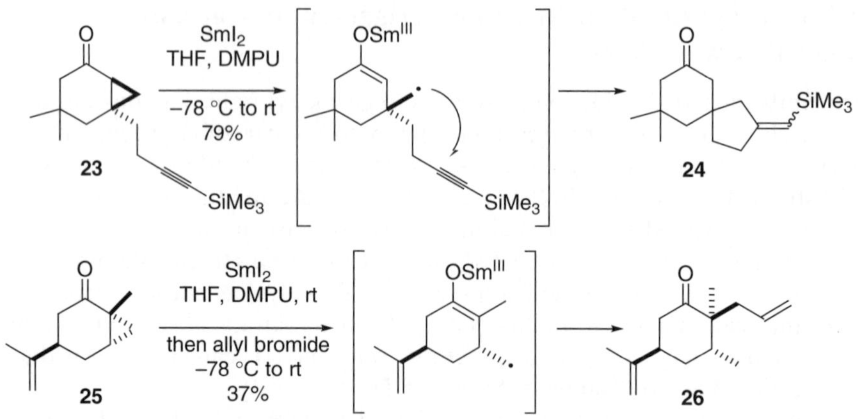

Scheme 4.13

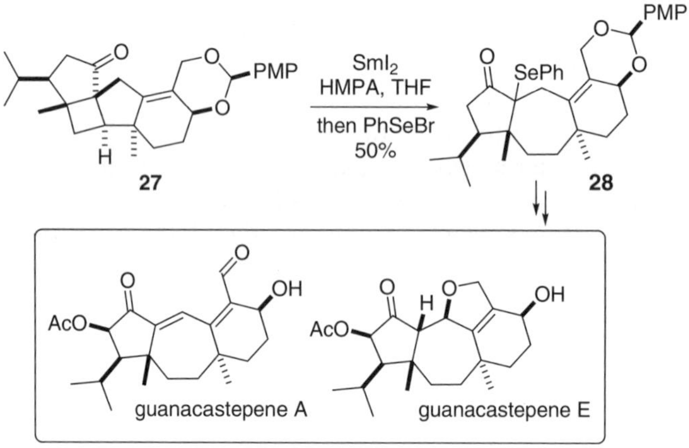

Scheme 4.14

treatment of **27** with the reagent gave selenide **28** after interception of the Sm(III) enolate intermediate with phenylselenyl bromide (Scheme 4.14).[24]

4.4 The Reduction of Carboxylic Acids, Esters and Amides with SmI_2

Although SmI_2 has been used for the partial reduction of α-heteroatom-substituted esters, lactones and amides (see Section 4.5), the reduction of simple, unfunctionalised substrates is not usually possible. It is now clear that in certain cases reduction using SmI_2 is possible provided that an activating additive is

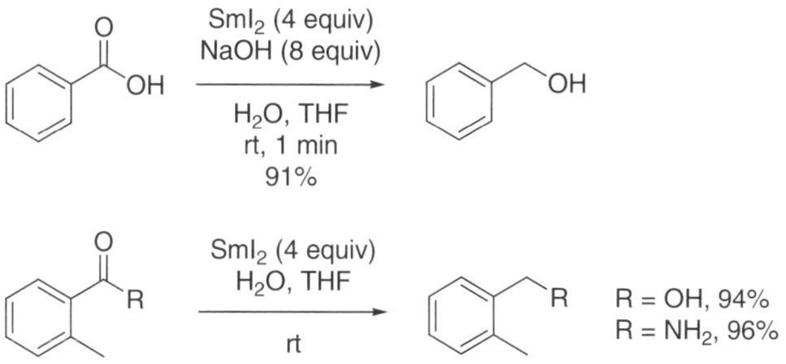

Scheme 4.15

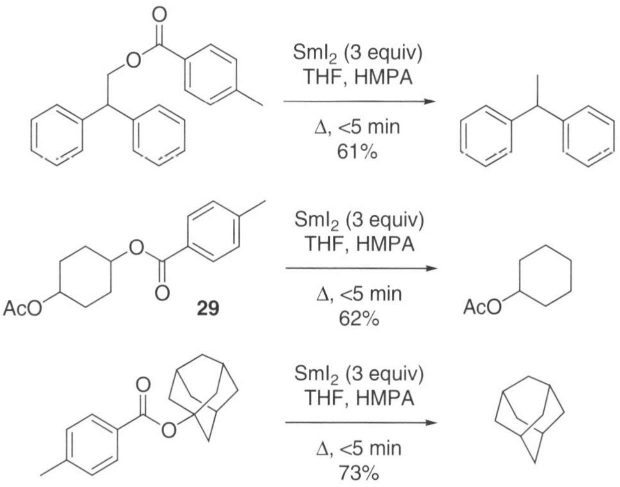

Scheme 4.16

also used. For example, Kudo reported the reduction of carboxylic acids with SmI_2, H_2O and NaOH[25] and the reduction of aryl acids, esters and amides using SmI_2 and H_2O (Scheme 4.15).[26] The H_2O cosolvent is known to activate SmI_2 (see Chapter 2, Section 2.2) and its inclusion is crucial to the success of these reductions.

Recently, Marko used the SmI_2-mediated reduction of aryl esters in a new deoxygenation protocol.[27] Reduction of primary, secondary and tertiary toluates using SmI_2–HMPA at reflux in THF (or THP) gave the corresponding deoxygenated products in good yield. The successful deoxygenation of **29** illustrates the selectivity of SmI_2 for aryl esters over alkyl esters (Scheme 4.16).[27]

Marko proposed that the reduction proceeds through ketyl radical anion **30** followed by radical fragmentation. As carrying out the reaction in THF-d_8 gave no labelled product, it would appear that the radical **31** is reduced to the alkylsamarium that is quenched by traces of proton source present in the reaction (Scheme 4.17).[27]

Procter showed recently that unfunctionalised lactones can be reduced using SmI_2–H_2O.[28] The reagent system is selective for the reduction of lactones over esters; furthermore, it displays complete ring size selectivity in that only six-membered lactones are converted to the corresponding diols. Experimental and computational studies suggest that the selectivity originates from the initial electron transfer to the lactone carbonyl to give a ketyl radical anion stabilised by an anomeric effect (Scheme 4.18).[28]

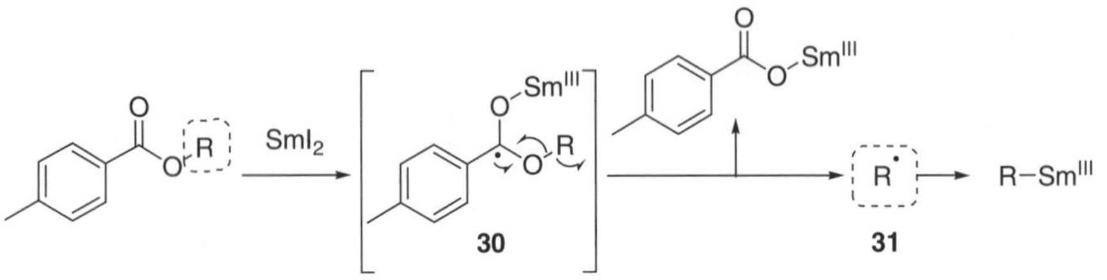

Scheme 4.17

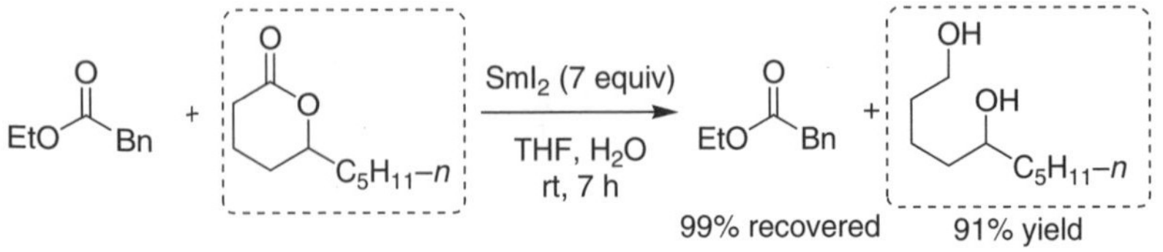

Scheme 4.18

Representative procedure.[28] To a stirred solution of SmI_2 in THF (6 equiv) was added distilled H_2O (25% of SmI_2 THF solution) and the colour of the solution changed from blue–green to red–black. A solution of the lactone (1 equiv) in THF was then added by cannula and the reaction mixture stirred for 3 h. The reaction was quenched by opening to air and the addition of aqueous saturated NH_4Cl, followed by extraction with Et_2O. The organic extracts were combined and dried (Na_2SO_4) and the solvent was removed *in vacuo*. The crude product was purified by column chromatography (EtOAc–hexane eluent).

4.5 The Reduction of α-Heteroatom-substituted Carbonyl Groups with SmI_2

The reduction of α-heteroatom-substituted carbonyl compounds to the parent carbonyl compound is one of the most common uses of SmI_2. Prior to the use of SmI_2, zinc, chromous ions and dissolving metal reduction had been used for the transformation. These methods often required acidic conditions and extended reaction times, thus limiting the range of functional groups tolerated. Using SmI_2, a range of heteroatom substituents can be removed efficiently under mild, neutral, electron-transfer conditions, in the presence of a range of other functional groups (Scheme 4.19).

The reduction of α-heteroatom-substituted carbonyl compounds is an excellent method for the generation of Sm(III) enolates that can be used in carbon–carbon bond-forming reactions (see Chapter 5, Section 5.5).

4.5.1 Mechanism of the Reduction of α-Heteroatom-substituted Carbonyl Groups with SmI_2

Two possible mechanisms can be envisaged for this reduction, both involving the formation and protonation of Sm(III) enolates. Molander proposed that α-halo ketones react with SmI_2 to give a ketyl radical **32** that is then quenched by the cosolvent.[29] A second reduction then gives carbanion **33** that undergoes β-elimination to produce the enol tautomer of the product ketone (Scheme 4.20).

As esters are not typically reduced by SmI_2, Molander proposed an alternative mechanism for the reduction of α-halo esters.[29] He proposed that

Scheme 4.19

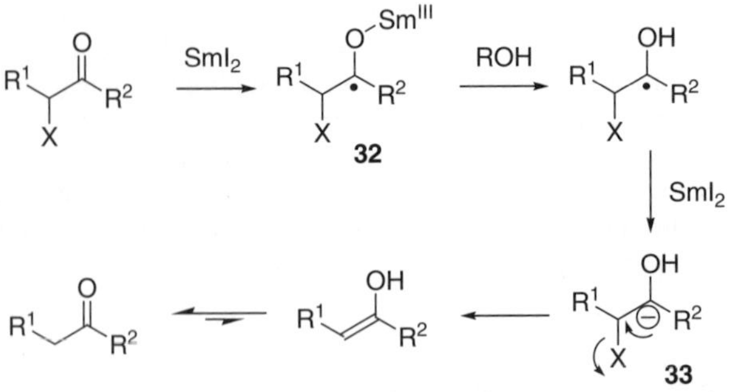

Scheme 4.20

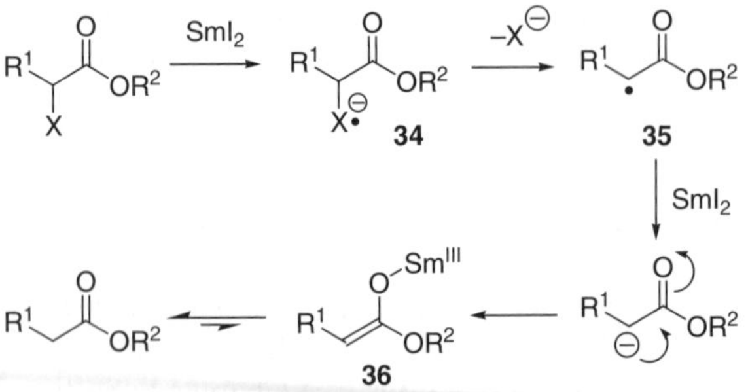

Scheme 4.21

electron transfer to the α-substituent generates a radical-anion intermediate **34** that undergoes fragmentation to give radical **35**. Reduction to the Sm(III) enolate **36** and protonation gives the parent ester (Scheme 4.21).

Alternatively, the electron-withdrawing properties of the α-heteroatom and its ability to coordinate SmI_2 may facilitate reduction of the ester carbonyl group and formation of ketyl radicals **37** (Scheme 4.22). Once the activating heteroatom group has been removed, no further ester reduction is possible. A similar mechanism is possible for the reduction of α-heteroatom-substituted amides.

4.5.2 Reduction of α-Heteroatom-substituted Ketones with SmI_2

In 1986, Molander carried out the first detailed study on the reduction of α-heteroatom-substituted carbonyl compounds with SmI_2.[29] It was shown that SmI_2 could reduce a range of α-oxygenated ketones under mild conditions to give the parent ketone in good yields and with no over-reduction of the ketone carbonyl (Scheme 4.23).

Representative procedure.[29] To a solution of SmI_2 in THF (2 equiv) at –78 °C was added the α-heteroatom-substituted carbonyl compound (1 equiv) in

MeOH–THF. The reaction mixture was stirred for 10 min, then allowed to warm to room temperature before being poured into saturated aqueous K_2CO_3. The aqueous layer was extracted with Et_2O and the combined organic layers were dried ($MgSO_4$) before concentration *in vacuo*. The crude products were then analysed by GC.

Molander also highlighted the chemoselectivity of the reaction: α-acetoxy iodoketone **38** was reduced with SmI_2 to give iodoketone **39** in 87% yield (Scheme 4.24).[29] In the same study, it was shown that α-halo, α-sulfanyl, α-sulfinyl and α-sulfonyl substrates underwent efficient reduction with SmI_2.[29]

The chemoselectivity of SmI_2 was further highlighted in Holton's synthesis of 10-deacetoxypaclitaxel (**41**).[30] 10-Deacetoxy derivatives of the anticancer agent paclitaxel (Taxol) have been found to have similar biological activity to the natural product. Previous routes to 10-deacetoxypaclitaxel derivatives proceeded in a number of steps and in low overall yield from paclitaxel (**40**). Holton showed that simple treatment of **40** with SmI_2 furnished the 10-deacetoxy derivative **41** in excellent yield with no requirement for protecting group chemistry (Scheme 4.25).[30]

The reduction of α,β-epoxy ketones using SmI_2 is also possible. A range of acyclic and cyclic epoxides with different substitution patterns underwent reduction to give the desired aldol products in good yields (Scheme 4.26).[31]

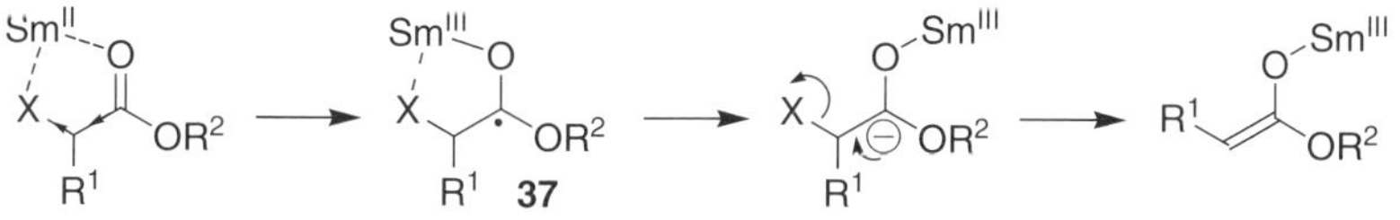

Scheme 4.22

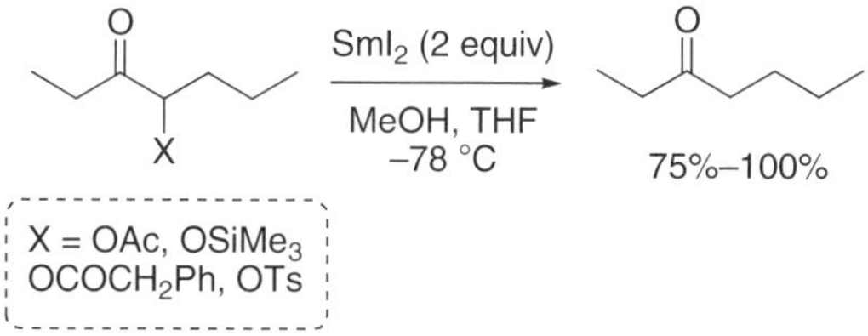

Scheme 4.23

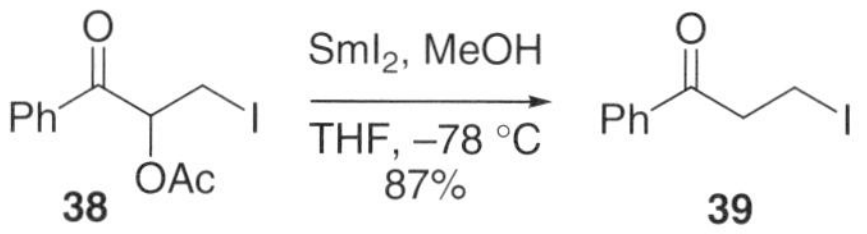

Scheme 4.24

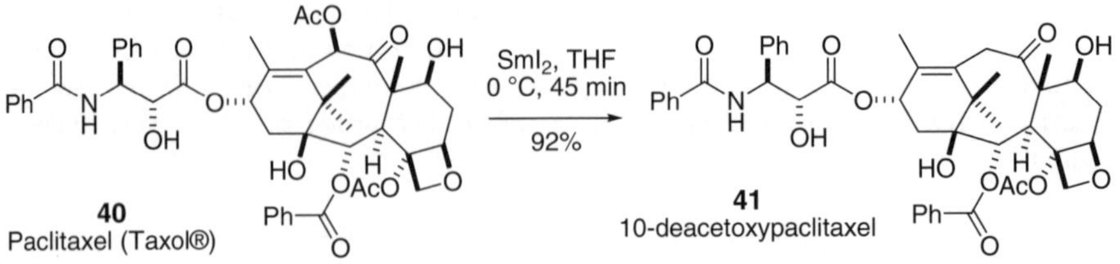

Scheme 4.25

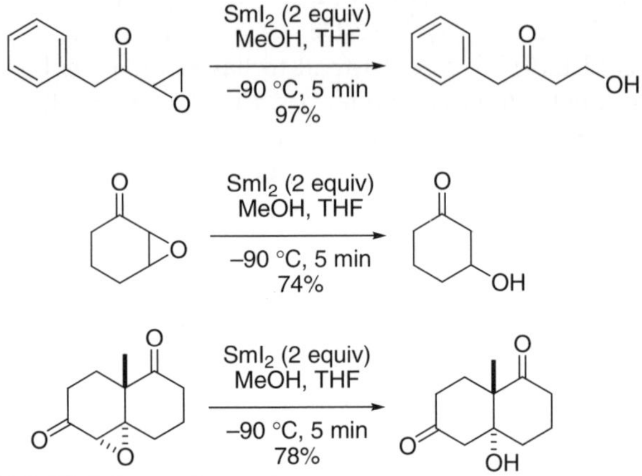

Scheme 4.26

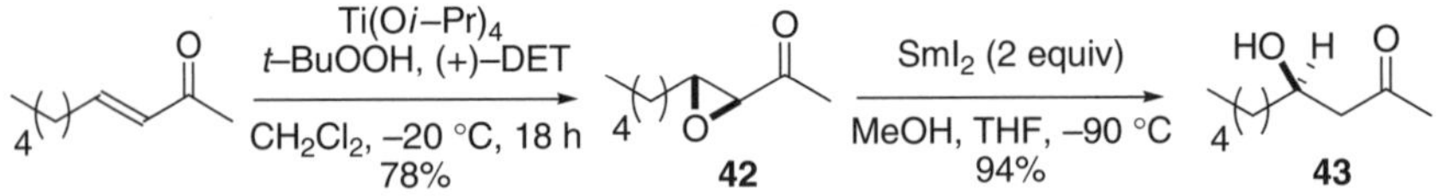

Scheme 4.27

Molander also showed that enantiomerically enriched α,β-epoxy ketones, such as **42**, prepared by Sharpless asymmetric epoxidation underwent efficient conversion to enantiomerically enriched β-hydroxy ketones **43** upon treatment with SmI_2 (Scheme 4.27).[31]

Molander also extended his study to include the SmI_2-mediated reduction of 2-acylaziridines, to give β-amino ketones (Scheme 4.28).[32]

4.5.3 Reduction of α-Heteroatom-substituted Esters with SmI_2

As stated previously, simple esters and lactones are not typically reduced with SmI_2. However, α-heteroatom-substituted esters and lactones can be partially

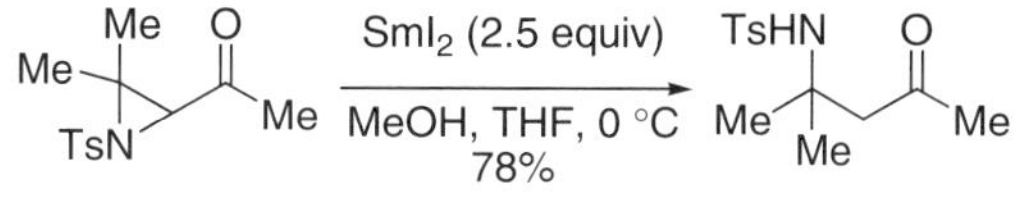

Me, Me, TsN (aziridine), O, Me — SmI_2 (2.5 equiv), MeOH, THF, 0 °C, 78% → TsHN, O, Me, Me, Me

Scheme 4.28

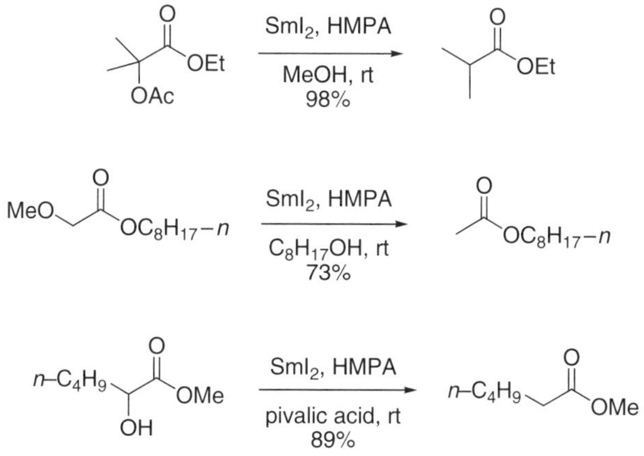

O, OEt, OAc — SmI_2, HMPA, MeOH, rt, 98% → O, OEt

MeO, O, OC_8H_{17}–*n* — SmI_2, HMPA, $C_8H_{17}OH$, rt, 73% → O, OC_8H_{17}–*n*

n–C_4H_9, O, OMe, OH — SmI_2, HMPA, pivalic acid, rt, 89% → *n*–C_4H_9, O, OMe

Scheme 4.29

reduced to the parent carbonyl compound. In 1989, Inanaga found that the deoxygenation of both α-acetoxy and α-methoxy esters proceeded well when HMPA was employed to increase the reduction potential of SmI_2. The reduction of α-hydroxy esters required a more acidic proton source and pivalic acid was used (Scheme 4.29).[33]

Representative procedure.[33] To a stirred solution of the α-oxygenated ester (1 equiv) and HMPA (5% of the SmI_2 THF solution) in MeOH under nitrogen was added a 0.1 M solution of SmI_2 in THF (2.5 equiv) at room temperature. The crude product was purified by column chromatography (EtOAc–hexane as eluent).

Inanaga also studied the reduction of α,β-epoxy esters with SmI_2 and found that a strong chelating agent such as dimethylaminoethanol (DMAE) or TMEDA was required to obtain high regioselectivity (Scheme 4.30).[34]

The addition of DMAE is thought to have two roles: first, it serves as a proton source for the Sm(III) enolate formed in the reaction, and second, it may act as a scavenging agent, removing the Lewis acidic Sm(III) salts from the reaction. Sm(III)-mediated opening of the epoxide led to the isolation of regioisomeric by-products. The use of DMAE minimises these side reactions. DMAE was also used as an additive in the reduction of aziridine-2-carboxylates; however, HMPA was not needed. A range of protected aziridines underwent efficient reduction with SmI_2 to give β-amino esters in good yield (Scheme 4.31).[32]

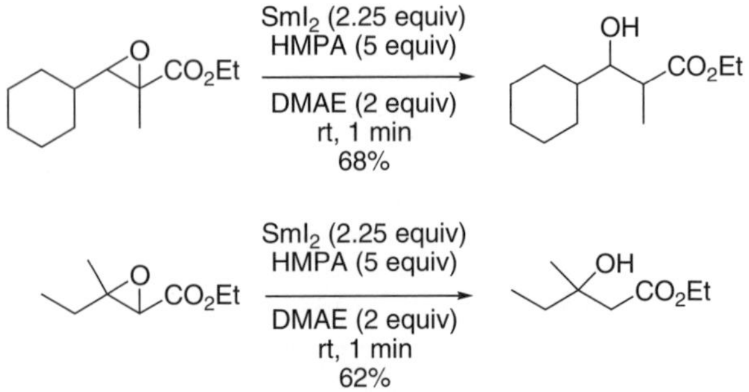

Scheme 4.30

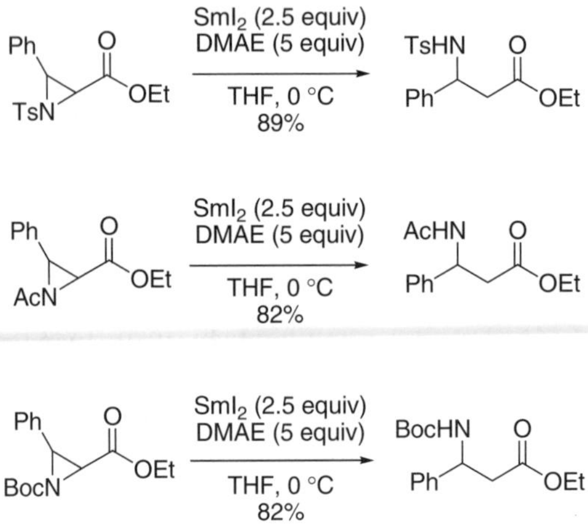

Scheme 4.31

4.5.4 Reduction of α-Heteroatom-substituted Amides with SmI_2

Fewer examples of the reduction of α-heteroatom-substituted amides with SmI_2 have been reported. Molander studied the reduction of *N*-tosylaziridine-2-carboxamides with SmI_2. As in the reduction of aziridine-2-carboxylates, the addition of DMAE was required for the reaction to proceed with high regiochemical control (Scheme 4.32).[32] It was also possible to reduce *N*-methoxy-*N*-methylaziridine-2-carboxamides with SmI_2 provided that the temperature and the amount of SmI_2 were carefully monitored to prevent N–O bond reduction (see Section 4.6.2).[32]

In 1999, Simpkins reported the first examples of the reduction of α-oxygenated amides.[35] Following Flowers' studies on the efficiency of LiCl as an additive in SmI_2 chemistry (see Chapter 2, Section 2.2.3),[36] the additive was

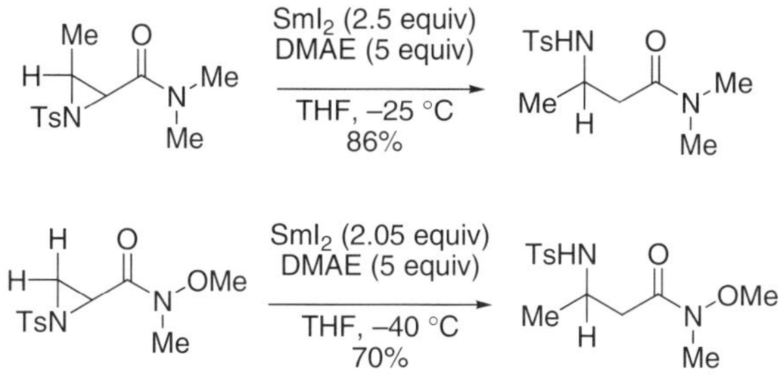

Scheme 4.32

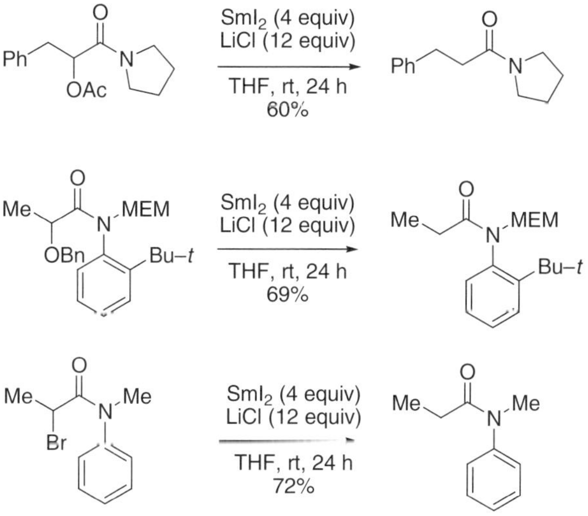

Scheme 4.33

found to enhance the relatively slow reduction of a range of α-acetoxy-, α-benzyloxy- and α-halo-substituted amides to provide good yields of the parent amides (Scheme 4.33).[35] Procter applied the reduction of α-heteroatom-substituted amides in a linker strategy for solid-phase and fluorous synthesis (see Chapter 7, Section 7.3).

Representative procedure.[35] To flame-dried LiCl in THF (20 equiv) was added SmI_2 in THF (3 equiv) and the solution was stirred for 30 min. The amide (1 equiv) in THF was then added and the reaction mixture stirred for 24 h. The mixture was poured into saturated aqueous Na_2SO_3 and extracted with EtOAc. The combined organic extracts were then washed with saturated aqueous Na_2SO_3 and brine and dried ($MgSO_4$) and the solvent was removed *in vacuo*. The crude product was purified by column chromatography (EtOAc–petroleum ether as eluent).

4.5.5 β-Elimination Reactions by the Reduction of α-Heteroatom-substituted Carbonyl Compounds with SmI_2

Concellón showed that the reduction of α-halo-β-hydroxy esters and amides, such as **44** and **45**, with SmI_2 gives α,β-unsaturated esters and amides, **46** and **47**, with high stereochemical control.[37,38] The diastereoselectivity of the process has been explained by invoking elimination through a six-membered chelate **48** (Scheme 4.34).

A number of other β-elimination processes have been developed by Concellón that are based on the reduction of α-heteroatom-substituted carbonyl compounds (Scheme 4.35).[7]

4.6 The Reduction of Nitrogen-containing Compounds with SmI_2

4.6.1 Reduction of the Nitro Group with SmI_2

In Kagan's early investigations, he showed that nitroarenes could be reduced to the corresponding anilines using SmI_2.[39] In 1991, Kende showed that primary, secondary or tertiary nitroalkanes could also be reduced using SmI_2 to the corresponding hydroxylamines or amines, depending on the amount of reagent used: the reaction of nitroalkanes and nitroarenes with 4 equiv of SmI_2, in the presence of MeOH and THF, gave the corresponding hydroxylamine, whereas the use of 6 equiv gave the amine (Scheme 4.36).[40]

Representative procedure.[40] To a stirred solution of SmI_2 in THF (4 equiv) was added rapidly a solution of the nitroalkane (1 equiv) in 2:1 THF–MeOH. The reaction mixture was stirred for 3 h, then poured into a 10% solution of Na_2SO_3 and extracted with EtOAc. The organic extracts were dried and concentrated *in vacuo*. The crude hydroxylamine product was purified by column chromatography (EtOAc as eluent).

Representative procedure.[40] To a stirred solution of SmI_2 in THF (6 equiv) was added a solution of the nitroalkane (1 equiv) in 2:1 THF–MeOH. The reaction mixture was stirred for 8 h, then poured into a 10% solution of Na_2SO_3 and extracted with EtOAc. The organic extracts were dried and concentrated *in vacuo* to give the crude amine product.

The mechanism of nitro group reduction by SmI_2 is thought to proceed by a pathway similar to that proposed for the dissolving metal reduction of such substrates: a series of four single-electron transfers from SmI_2 followed by proton transfers generate hydroxylamine intermediates that can then be reduced further. Sm(III) salts generated during the reaction presumably facilitate the departure of a hydroxyl leaving group (Scheme 4.37).

In 2002, Brady reported a mechanistic study of the Sm(II)-mediated reduction of nitroarenes to anilines that provides a more detailed picture of nitro group reduction.[41] The use of $Sm[N(TMS)_2]_2$ in place of SmI_2 allowed a

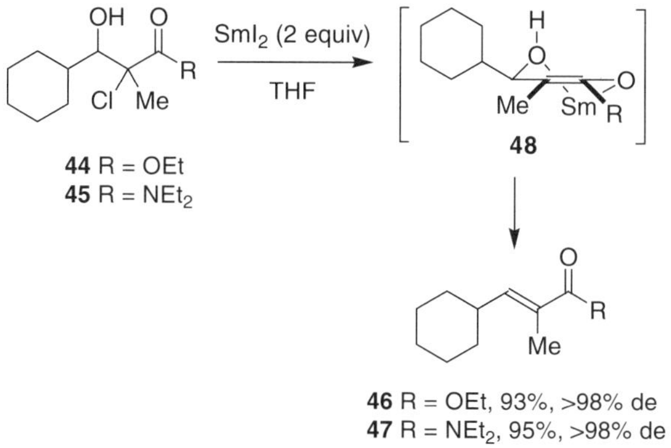

Scheme 4.34

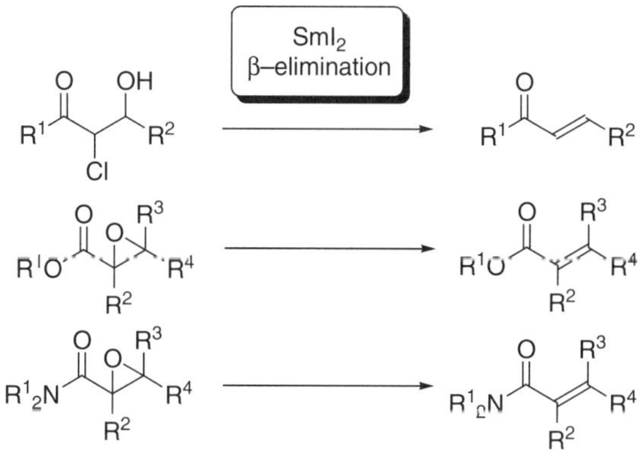

Scheme 4.35

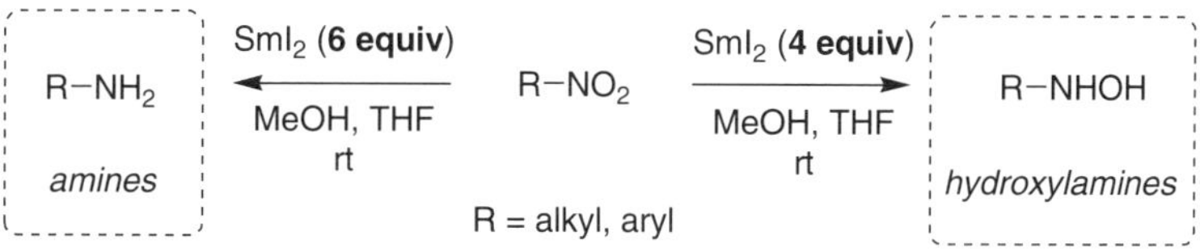

Scheme 4.36

detailed study of the intermediates formed in the reaction by ^{1}H NMR spectroscopy and X-ray crystallography.[41] Treatment of nitrobenzene with 2 equiv of the Sm(II) complex formed nitrosobenzene **49** and a further 2 equiv generated azobenzene intermediate **50**. Further reduction formed the azo-bridged bimetallic species **51** and then **52** before hydrolysis formed aniline (Scheme 4.38).[41]

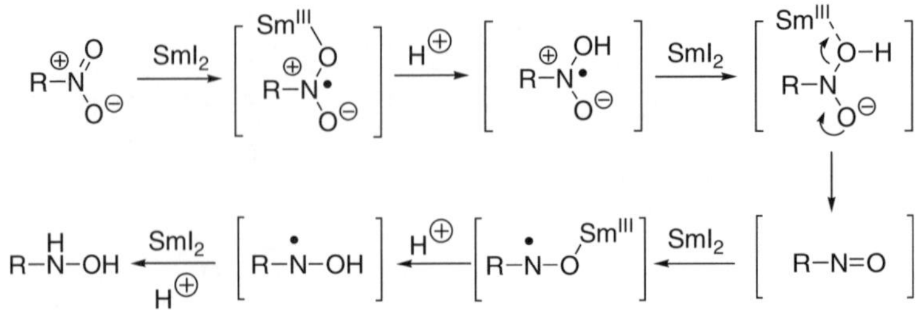

Scheme 4.37

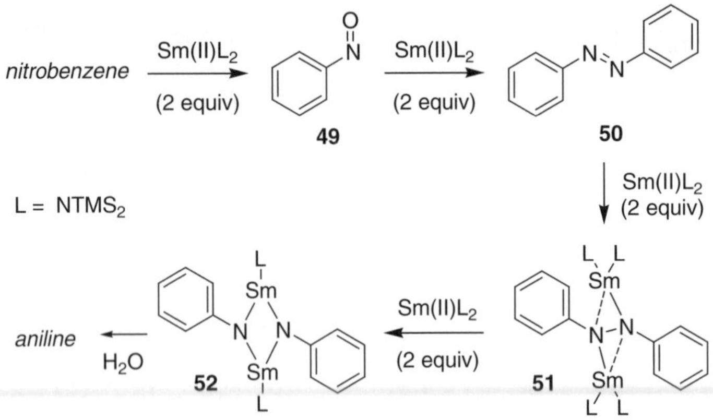

Scheme 4.38

4.6.2 Reduction of N–O Bonds with SmI_2

SmI_2 can reduce N–O bonds in many other substrates. In 1995, Keck studied the reduction of the N–O bond in amide **53**. Whereas reducing systems including Al(Hg), Na(Hg) and hydrogenolysis proved unsuccessful, treatment of **53** with SmI_2 resulted in smooth N–O bond reduction to give **54** in 91% yield (Scheme 4.39).[42]

Related reductions using SmI_2 have also been reported. For example, Zhang described the rapid reduction of pyridine *N*-oxides and azobenzenes using SmI_2 with MeOH,[43] whereas Natale used SmI_2 to cleave the N–O bond in isoxazoles.[44]

Carreira utilised the SmI_2 reduction of the N–O bond in isoxazolines as a key step in the diastereoselective construction of β-hydroxycarbonyl motifs in target synthesis.[45,46] For example, conjugated isoxazolines, such as **55**, formed by highly diastereoselective, intermolecular nitrone–alkene cycloadditions, are reduced to the corresponding β′-hydroxy α,β-unsaturated ketones **56** using SmI_2. The product imine is hydrolysed during work-up with H_2O and $B(OH)_3$ (Scheme 4.40).[45] Previously, such reductions had been carried out using Raney

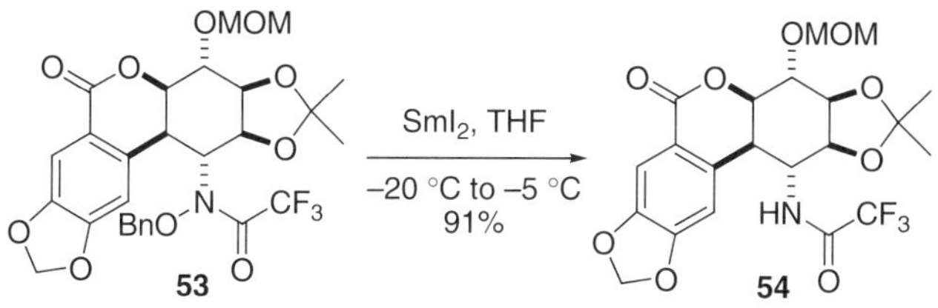

Scheme 4.39

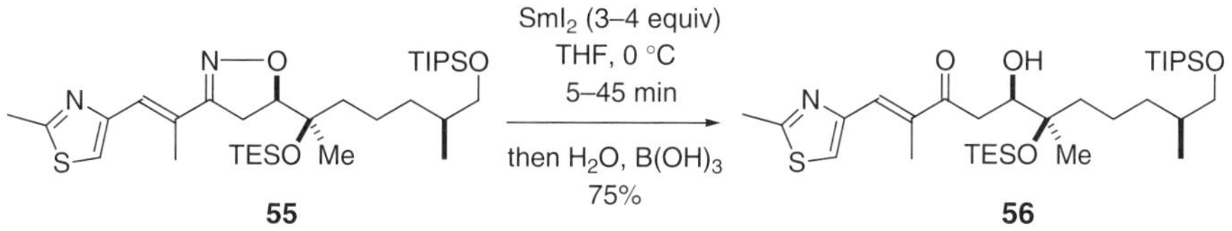

Scheme 4.40

Ni, $TiCl_3$ and $Mo(CO)_6$. The SmI_2 reduction has been shown to be selective for conjugated isoxazolines in the presence of unconjugated isoxazolines.[46]

4.6.3 Reduction of N–N Bonds with SmI_2

Since Kagan's first report of N–N bond reduction using SmI_2,[39] the reagent is now routinely used for the transformation. The reaction has been used widely in asymmetric synthesis to reduce the products of asymmetric additions to, or reductions of,[47] hydrazone derivatives. For example, Enders reported the nucleophilic addition of alkyllithiums to trifluoroacetaldehyde SAMP and RAMP hydrazones in an asymmetric approach to α-trifluoromethyl-substituted amines.[48] After activation of the adducts by benzoylation, SmI_2-mediated N–N bond cleavage proceeded in high yield (Scheme 4.41).

Selected further examples of the SmI_2-mediated reduction of activated N–N bonds include reduction of adducts resulting from asymmetric allylation[49] and organocatalytic α-amination[50] (Scheme 4.42).

4.7 The Reduction of Electron-deficient Alkenes with SmI_2

In the early 1990s, Inanaga[51] and Alper[52] independently reported the conjugate reduction of α,β-unsaturated esters and amides with SmI_2. Whereas Inanaga used SmI_2 and *N,N*-dimethylacetamide (DMA) for the transformation,[51] Alper employed the SmI_2–HMPA reagent system.[52] Alper also reported the rapid conjugate reduction of α,β-unsaturated mono- and diacids using SmI_2–HMPA (Scheme 4.43).[52]

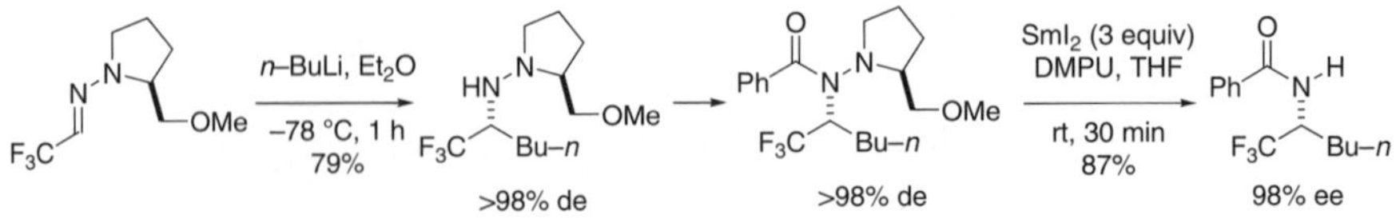

Scheme 4.41

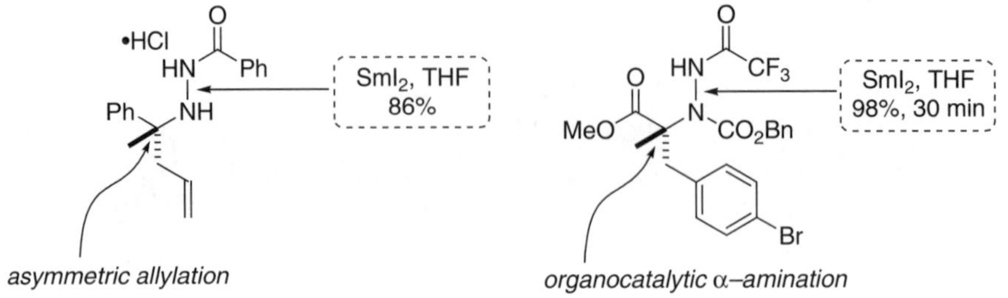

Scheme 4.42

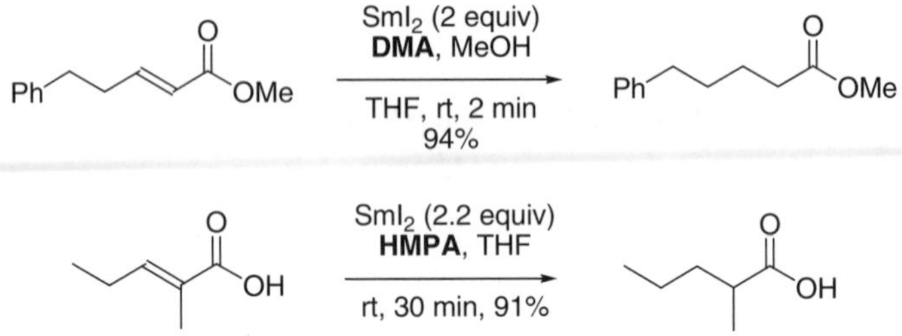

Scheme 4.43

Representative procedure.[52] To a stirred solution of the α,β-unsaturated ester (1 equiv) in dry THF was added a 0.1 M solution of SmI_2 in THF (2 equiv) followed by HMPA (9 equiv) and the reaction mixture was stirred for 30 min. The mixture was diluted with H_2O and extracted with Et_2O. The organic extracts were dried ($MgSO_4$) and concentrated *in vacuo* to give the crude product.

Fukuzumi and Otera studied the reduction of electron-deficient alkenes in lactones and cyclic enones using SmI_2.[53] More recently, the use of Hilmersons' SmI_2–H_2O-amine system was shown to be effective for the selective reduction of α,β-unsaturated esters and conjugated double and triple bonds (Scheme 4.44).[54,55]

Keck utilised SmI_2 to reduce an electron-deficient alkene in an approach to the marine natural product dolabelide B.[56] Treatment of α,β-unsaturated ester **57** with SmI_2 in the presence of MeOH and DMA gave **58** in moderate yield (Scheme 4.45).[56]

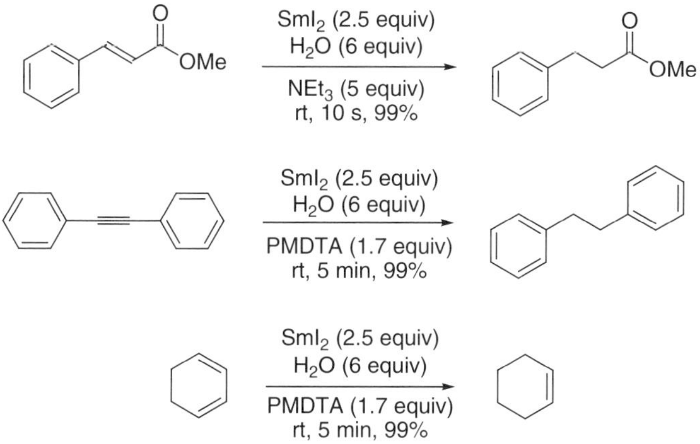
O
OMe
SmI_2 (2.5 equiv)
H_2O (6 equiv)
NEt_3 (5 equiv)
rt, 10 s, 99%
O
OMe
SmI_2 (2.5 equiv)
H_2O (6 equiv)
PMDTA (1.7 equiv)
rt, 5 min, 99%
SmI_2 (2.5 equiv)
H_2O (6 equiv)
PMDTA (1.7 equiv)
rt, 5 min, 99%

Scheme 4.44

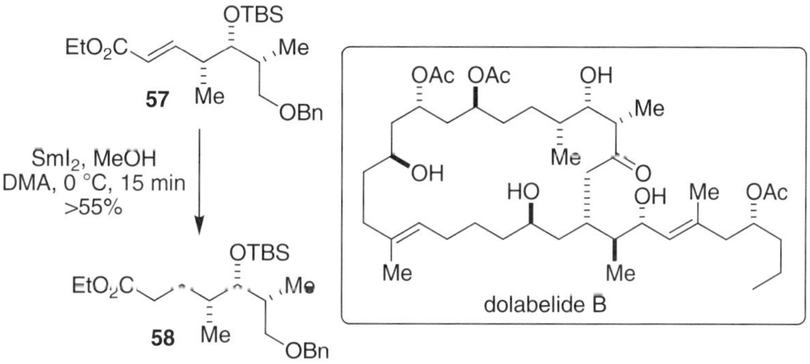
OTBS
EtO_2C
Me
57
Me
OBn
SmI_2, MeOH
DMA, 0 °C, 15 min
>55%
OTBS
EtO_2C
Me
58
Me
OBn
OAc OAc
OH
Me
Me
OH
O
HO
OH
Me
OAc
Me
Me
dolabelide B

Scheme 4.45

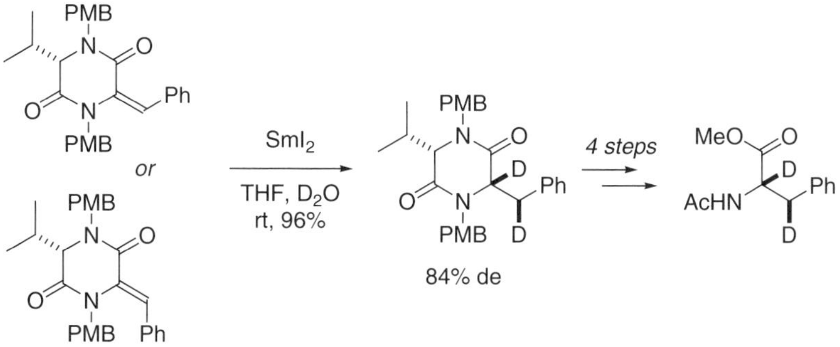
PMB
N
O
Ph
O
N
PMB
or
PMB
N
O
O
N
PMB Ph
SmI_2
THF, D_2O
rt, 96%
PMB
N
O
D
Ph
O
N
PMB D
84% de
4 steps
MeO
O
D
AcHN
Ph
D

Scheme 4.46

Davies used SmI_2 and D_2O to carry out a highly diastereoselective conjugate reduction of benzylidene diketopiperazine templates *en route* to (2*S*,3*R*)-dideuteriophenylalanine.[57] The new stereocentres arise from diastereoselective protonation of a Sm(III) enolate and the protonation of a stereodefined organosamarium intermediate (Scheme 4.46).[58]

Procter utilised the conjugate reduction of α,β-unsaturated esters to generate Sm(III) enolates for use in stereoselective aldol cyclisations (Chapter 5, Section 5.5).

4.7.1 Mechanism of the Electron-deficient Alkene Reduction with SmI_2

The reduction of electron-deficient alkenes is thought to proceed through a radical enolate species **59**. Further reduction generates dianionic species **60** that is then protonated to give the product of reduction (Scheme 4.47).

Fukuzumi and Otera showed that substituents at the β-position increased the rate of reduction of the electron-deficient alkene. This supports the above mechanism as the more substituted the β-carbon centre, the more stable is the β-radical and the more facile the reduction (Scheme 4.48).[53]

4.8 Miscellaneous Functional Group Reductions with SmI_2

4.8.1 Reduction of Alkylsulfonates

There is limited precedent for the reduction of alkyltosylates to alkanes using SmI_2. It is likely that the mechanism of reduction involves *in situ* conversion of the alkyl tosylate to the alkyl iodide followed by reduction of the iodide.[1]

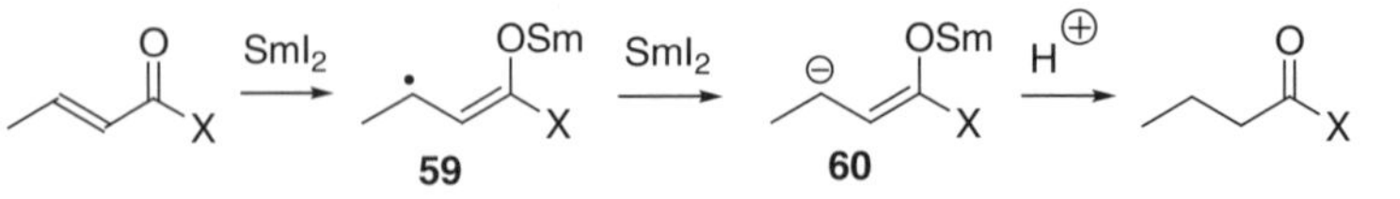

Scheme 4.47

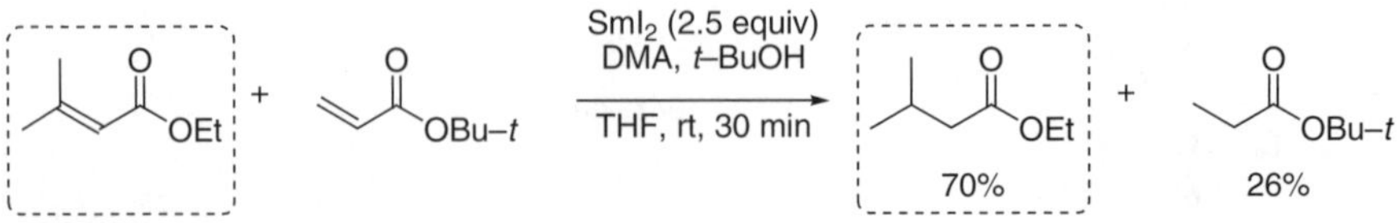

Scheme 4.48

4.8.2 Reduction of Sulfoxides and Sulfones with SmI_2

Kagan reported that the reduction of sulfoxides to sulfides using SmI_2 in THF is possible but requires long reaction times.[1] The addition of HMPA greatly accelerates the reduction of sulfoxides and dialkyl, diaryl and aryl alkyl sulfoxides were converted to the corresponding sulfides on treatment with a large excess of SmI_2 (Scheme 4.49).[59]

In his early studies, Kagan showed that SmI_2 can be used for the deoxygenation of epoxides.[1] The deoxygenation of phosphine oxides and tin oxides is also effectively achieved using SmI_2 provided that HMPA is used as an additive.[59]

The reduction of sulfones with SmI_2 can result in deoxygenation, to give the parent sulfide,[59] or carbon–sulfur bond cleavage, depending on the substrate and the reaction conditions employed.[60] For example, alkyl and alkenyl phenyl sulfones are reductively cleaved by SmI_2 in the presence of HMPA. In the reduction of alkenyl sulfones, competing reduction of the electron-deficient alkene is not observed (Scheme 4.50).[60]

Gin utilised the reduction of an alkyl phenyl sulfone in a synthesis of the anti-leukaemia natural product (–)-deoxyharringtonine:[61] Treatment of complex sulfone **61** with SmI_2-HMPA gave **62** in 74% yield (Scheme 4.51).

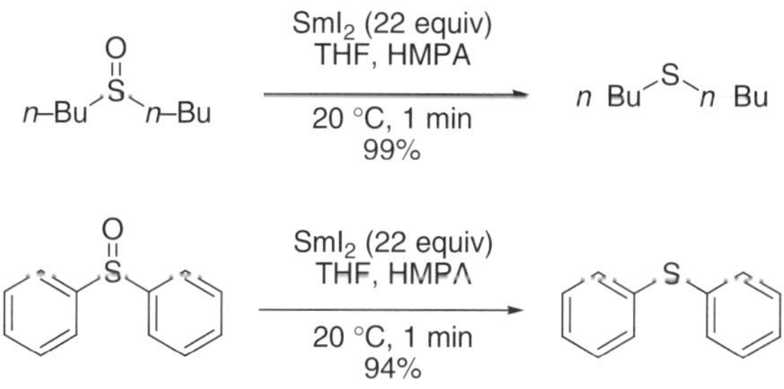

SmI_2 (22 equiv), THF, HMPA; 20 °C, 1 min; 99% (n-Bu–S(O)–n-Bu → n Bu–S–n Bu)

SmI_2 (22 equiv), THF, HMPA; 20 °C, 1 min; 94%

Scheme 4.49

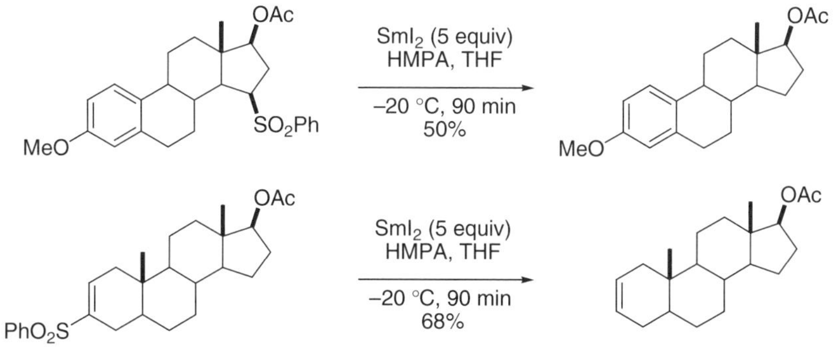

SmI_2 (5 equiv), HMPA, THF; –20 °C, 90 min; 50% (OAc, SO_2Ph, MeO → OAc, MeO)

SmI_2 (5 equiv), HMPA, THF; –20 °C, 90 min; 68% (PhO_2S, OAc → OAc)

Scheme 4.50

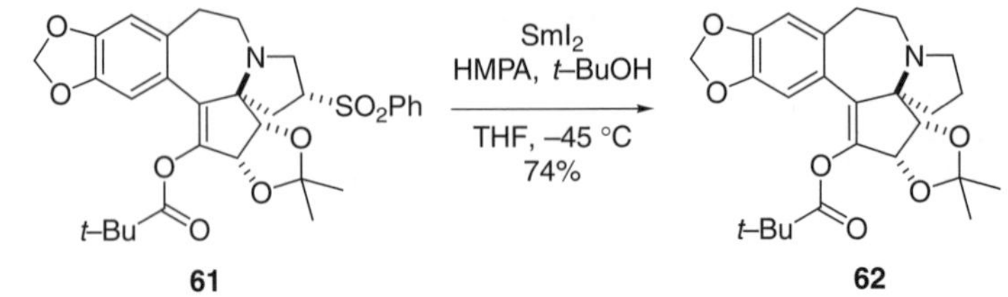

Scheme 4.51

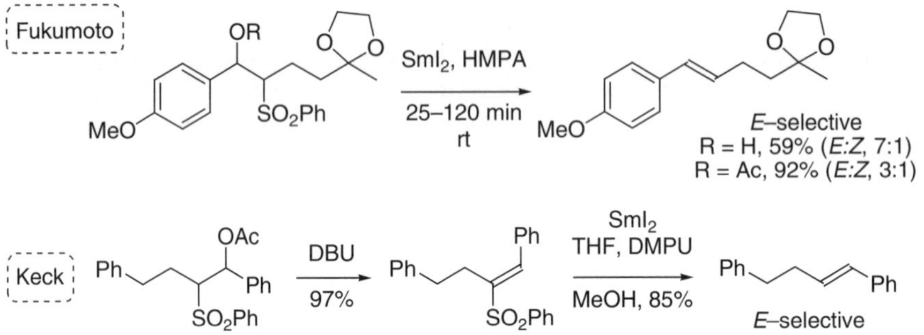

Scheme 4.52

Representative procedure.[61] To a stirred solution of *t*-BuOH (10 equiv) was added the sulfone **61** (1 equiv) and the solution was cooled to −78 °C. A solution of SmI_2 in THF (5 equiv) was then added and the reaction mixture stirred for 10 min followed by the addition of HMPA (2.5 equiv). The reaction mixture was stirred at −78 °C for 1 h, then at −45 °C for 2 h before warming to room temperature, diluting with H_2O and extraction with EtOAc. The combined organic extracts were dried ($MgSO_4$) and the solvent was removed *in vacuo*. Crude product **62** was purified by column chromatography (EtOAc–benzene eluent).

One of the most important applications of sulfone reduction using SmI_2 is in the modified Julia–Lythgoe olefination. In 1990, Kende reported the use of SmI_2 in the reductive elimination of β-hydroxy imidazolyl sulfones to give (*E*)-alkenes.[62] Subsequently, Fukumoto[63] and Keck[64] independently reported the use of SmI_2 in olefination processes involving phenyl sulfones. In Fukumoto's system, SmI_2–HMPA was used to reduce both β-hydroxy and β-acetoxy sulfones,[63] whereas Keck used SmI_2–DMPU to reduce intermediate vinyl sulfones.[64] Both approaches are selective for the formation of (*E*)-alkenes. The use of SmI_2 shows considerable advantages over the use of Na–Hg in the classical olefination (Scheme 4.52).

More recently, Markó has used SmI_2 to reduce β-benzoyloxy sulfones[65] and sulfoxides[66] in modified Julia–Lythgoe olefinations.

The accelerating effect of HMPA on the reduction of the S–O bond has allowed SmI_2 to be used in excess for the deoxygenation of diaryl sulfones. Only moderate yields are obtained for the reduction of dialkyl sulfones.[59] As discussed above, the reduction of aryl alkyl sulfones can be accompanied by reductive cleavage.[60]

4.8.3 Peroxide Reduction Using SmI_2

Cyclic peroxides can be reduced to the corresponding diols by SmI_2.[67] The reduction has been shown to occur in the presence of alkenes, alcohols and primary alkyl tosylates (Scheme 4.53).[68]

4.8.4 Reductions Using SmI_2 and Pd

SmI_2 has also been employed as a stoichiometric reductant in deoxygenation processes catalysed by palladium complexes.[69] For example, allylic acetates can be reduced by treatment with SmI_2 and 2-propanol in the presence of a palladium(0) catalyst. Unfortunately, this reduction often generates regioisomeric and stereoisomeric mixtures of alkenes (Scheme 4.54).[69]

The reductive cleavage of propargyl γ-lactones using SmI_2 and a palladium(0) has been used to access allenes (Scheme 4.55).[70]

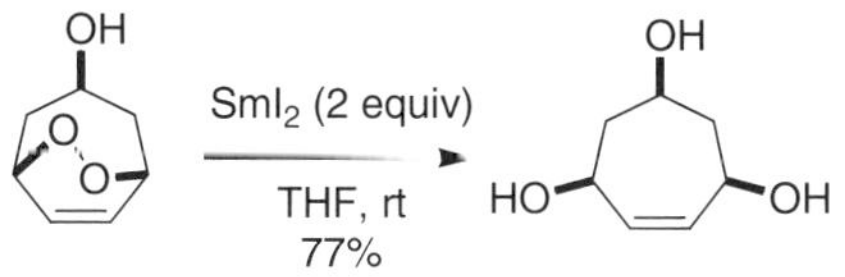

OH

SmI_2 (2 equiv)

THF, rt

77%

HO OH OH

Scheme 4.53

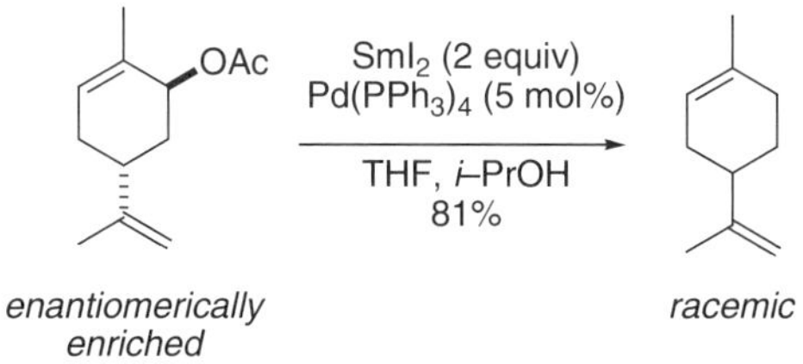

OAc

SmI_2 (2 equiv)

$Pd(PPh_3)_4$ (5 mol%)

THF, *i*-PrOH

81%

enantiomerically enriched

racemic

Scheme 4.54

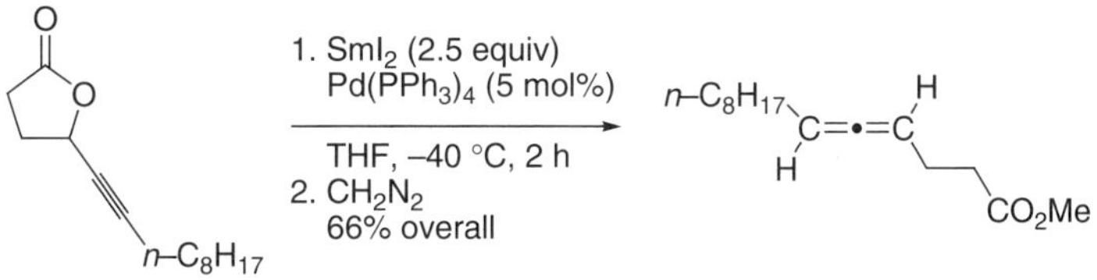

1. SmI_2 (2.5 equiv)
$Pd(PPh_3)_4$ (5 mol%)
THF, −40 °C, 2 h
2. CH_2N_2
66% overall

n-C_8H_{17}

n-C_8H_{17} H C=•=C H CO_2Me

Scheme 4.55

4.9 Conclusions

Although SmI_2 is perhaps best known for the powerful carbon–carbon bond-forming reactions that it initiates, it is important to remember that these transformations are initiated by the reduction of functional groups. The ability to manipulate a particular functional group in the presence of others is an important characteristic of the reagent and SmI_2 is now well established as 'the reagent of choice' for many functional group reductions. As the reductive range of SmI_2 is extended through a better understanding of activation through the use of additives, new functional group chemistry will be possible and exciting new selectivities will be discovered.

References

1. P. Girard, J.-L. Namy and H. B. Kagan, *J. Am. Chem. Soc.*, 1980, **102**, 2693.
2. J. Inanaga, M. Ishikawa and M. Yamaguchi, *Chem. Lett.*, 1987, 1485.
3. H. B. Kagan, J. L. Namy and P. Girard, *Tetrahedron*, 1981, **37**, 175.
4. D. P. Curran, T. L. Fevig, C. P. Jasperse and M. J. Totleben, *Synlett*, 1992, 943.
5. E. Hasegawa and D. P. Curran, *Tetrahedron Lett.*, 1993, **34**, 1717.
6. T. P. Ananthanarayan, T. Gallagher and P. Magnus, *J Chem. Soc. Chem. Commun.*, 1982, 709.
7. J. M. Concellón and H. Rodríguez-Solla, *Chem. Soc. Rev.*, 2004, **33**, 599.
8. J. M. Concellón, P. L. Bernad and J. A. Pérez-Andrés, *Angew. Chem. Int. Ed.*, 1999, **38**, 2384.
9. J. M. Concellón, P. L. Bernad and E. Bardales, *Org. Lett.*, 2001, **3**, 937.
10. (a) V. Rautenstrauch, B. Willhalm and W. Thommen, *Helv. Chim. Acta*, 1981, **64**, 2109; (b) Y.-D. Wu and K. N. Houk, *J. Am. Chem. Soc.*, 1992, **114**, 1656.
11. P. R. Chopade, E. Prasad and R. A. Flowers, II, *J. Am. Chem. Soc.*, 2004, **126**, 44.
12. A. K. Singh, R. K. Bakshi and E. J. Corey, *J. Am. Chem. Soc.*, 1987, **109**, 6187.
13. G. E. Keck, C. A. Wager, T. Sell and T. T. Wager, *J. Org. Chem.*, 1999, **64**, 2172.
14. G. E. Keck and C. A. Wager, *Org. Lett.*, 2000, **2**, 2307.
15. P. Chopade, T. A. Davis, E. Prasad and R. A. Flowers, II, *Org. Lett.*, 2004, **6**, 2685.
16. T. A. Davis, P. R. Chopade, G. Hilmersson and R. A. Flowers, II, *Org. Lett.*, 2005, **7**, 119.
17. G. E. Keck and A. P. Truong, *Org. Lett.*, 2002, **4**, 3131.
18. G. E. Keck, R. L. Giles, V. J. Cee, C. A. Wager, T. Yu and M. B. Kraft, *J. Org. Chem.*, 2008, **73**, 9675.
19. L.-L. Huang, M.-H. Xu and G.-Q. Lin, *J. Am. Chem. Soc.*, 2006, **128**, 5624.
20. G. A. Molander and J. A. McKie, *J. Org. Chem.*, 1991, **56**, 4112.

21. F. McKerlie, I. M. Rudkin, G. Wynne and D. J. Procter, *Org. Biomol. Chem.*, 2005, **3**, 2805.
22. H. Kusama, R. Hara, S. Kawahara, T. Nishimori, H. Kashima, N. Nakamura, K. Morihira and I. Kuwajima, *J. Am. Chem. Soc.*, 2000, **122**, 3811.
23. R. A. Batey, J. D. Harling and W. B. Motherwell, *Tetrahedron*, 1996, **52**, 11421.
24. W. D. Shipe and E. J. Sorensen, *J. Am. Chem. Soc.*, 2006, **128**, 7025.
25. Y. Kamochi and T. Kudo, *Chem. Lett.*, 1991, 893.
26. Y. Kamochi and T. Kudo, *Chem. Lett.*, 1993, 1495.
27. K. Lam and I. E. Markó, *Org. Lett.*, 2008, **10**, 2773.
28. L. A. Duffy, H. Matsubara and D. J. Procter, *J. Am. Chem. Soc.*, 2008, **130**, 1136.
29. G. A. Molander and G. Hahn, *J. Org. Chem.*, 1986, **51**, 1135.
30. R. A. Holton, C. Somoza and K.-B. Chai, *Tetrahedron Lett.*, 1994, **35**, 1665.
31. G. A. Molander and G. Hahn, *J. Org. Chem.*, 1986, **51**, 2596.
32. G. A. Molander and P. J. Stengel, *Tetrahedron*, 1997, **53**, 8887.
33. K. Kusuda, J. Inanaga and M. Yamaguchi, *Tetrahedron Lett.*, 1989, **30**, 2945.
34. K. Otsubo, J. Inanaga and M. Yamaguchi, *Tetrahedron Lett.*, 1987, **28**, 4437.
35. A. D. Hughes, D. A. Price and N. S. Simpkins, *J. Chem. Soc., Perkin Trans.*, 1999, 1295.
36. J. R. Fuchs, M. L. Mitchell, M. Shabangi and R. A. Flowers, *Tetrahedron Lett.*, 1997, **38**, 8157.
37. J. M. Concellón, J. A. Pérez-Andrés and H. Rodríguez-Solla, *Angew. Chem. Int. Ed.*, 2000, **39**, 2773.
38. J. M. Concellón, J. A. Pérez-Andrés and H. Rodríguez-Solla, *Chem. Eur. J.*, 2001, **7**, 3062.
39. J. Souppe, L. Danon, J. L. Namy and H. B. Kagan, *J. Organomet. Chem.*, 1983, **250**, 227.
40. A. S. Kende and J. S. Mendoza, *Tetrahedron Lett.*, 1991, **32**, 1699.
41. E. D. Brady, D. L. Clark, D. W. Keogh, B. L. Scott and J. G. Watkin, *J. Am. Chem. Soc.*, 2002, **124**, 7007.
42. G. E. Keck, S. F. McHardy and T. T. Wager, *Tetrahedron Lett.*, 1995, **36**, 7419.
43. Y. Zhang and R. Lin, *Synth. Commun.*, 1987, **17**, 329.
44. N. R. Natale, *Tetrahedron*, 1982, **23**, 5009.
45. J. W. Bode and E. M. Carreira, *Org. Lett.*, 2001, **3**, 1587.
46. L. D. Fader and E. M. Carreira, *Org. Lett.*, 2004, **6**, 2485.
47. M. J. Burk and J. E. Feaster, *J. Am. Chem. Soc.*, 1992, **114**, 6266.
48. D. Enders and K. Funabiki, *Org. Lett.*, 2001, **3**, 1575.
49. R. Berger, K. Duff and J. L. Leighton, *J. Am. Chem. Soc.*, 2004, **126**, 5686.
50. N. S. Chowdari and C. F. Barbas, III, *Org. Lett.*, 2005, **7**, 867.

51. J. Inanaga, Y. Handa, T. Tabuchi, K. Otsubo, M. Yamaguchi and T. Hanamoto, *Tetrahedron Lett.*, 1991, **32**, 6557.
52. A. Cabrera and H. Alper, *Tetrahedron Lett.*, 1992, **33**, 5007.
53. Y. Fujita, S. Fukuzumi and J. Otera, *Tetrahedron Lett.*, 1997, **38**, 2121.
54. A. Dahlen and G. Hilmersson, *Tetrahedron Lett.*, 2003, **44**, 2661.
55. A. Dahlen and G. Hilmersson, *Chem. Eur. J.*, 2003, **9**, 1123.
56. G. E. Keck and M. D. McLaws, *Tetrahedron Lett.*, 2005, **46**, 4911.
57. S. G. Davies, H. Rodríguez-Solla, J. A. Tamayo, C. Garner and A. D. Smith, *Chem. Commun.*, 2004, 2502.
58. S. G. Davies, H. Rodríguez-Solla, J. A. Tamayo, A. R. Cowley, C. Concellón, A. C. Garner, A. L. Parkes and A. D. Smith, *Org. Biomol. Chem.*, 2005, **3**, 1435.
59. Y. Handa, J. Inanaga and M. Yamaguchi, *J. Chem. Soc., Chem. Commun.*, 1989, 298.
60. H. Kunzer, M. Stahnke, G. Sauer and R. Wiechert, *Tetrahedron Lett.*, 1991, **32**, 1949.
61. J. D. Eckelbarger, J. T. Wilmot and D. Y. Gin, *J. Am. Chem. Soc.*, 2006, **128**, 10370.
62. A. S. Kende and J. S. Mendoza, *Tetrahedron Lett.*, 1990, **31**, 7105.
63. M. Ihara, S. Suzuki, T. Taniguchi, Y. Tokunaga and K. Fukumoto, *Synlett*, 1994, 859.
64. G. E. Keck, K. A. Savin and M. A. Weglarz, *J. Org. Chem.*, 1995, **60**, 3194.
65. I. E. Markó, F. Murphy, L. Kumps, A. Ates, R. Touillaux, D. Craig, S. Carballares and S. Dolan, *Tetrahedron*, 2001, **57**, 2609.
66. J. Pospísil, T. Pospísil and I. E. Markó, *Org. Lett.*, 2005, **7**, 2373.
67. C. R. Johnson and C. H. Senanayake, *J. Org. Chem.*, 1989, **54**, 735.
68. G. H. Posner and C. H. Oh, *J. Am. Chem. Soc.*, 1992, **114**, 6266.
69. T. Tabuchi, J. Inanaga and M. Yamaguchi, *Tetrahedron Lett.*, 1986, **27**, 601.
70. T. Tabuchi, J. Inanaga and M. Yamaguchi, *Tetrahedron Lett.*, 1986, **27**, 5237.

CHAPTER 5

Carbon–Carbon Bond-forming Reactions Using SmI_2

5.1 Pinacol Couplings

5.1.1 Introduction

Since the discovery of the pinacol coupling reaction over 150 years ago by Wilhelm Rudolf Fittig,[1] this reaction has become an important synthetic tool for the construction of carbon–carbon bonds.[2–6] The reductive coupling reaction between two carbonyl compounds represents a valuable alternative to the osmium tetraoxide-mediated dihydroxylation of alkenes for generating 1,2-diols (Scheme 5.1). Typically, low-valent metals or complexes thereof can promote these reactions and it is well accepted that metal ketyl radical anions are intermediates in these reactions. Due to the ease with which the reagent is prepared and used and the characteristically high yields of pinacolic products, SmI_2 has become one of the most popular reagents for mediating this class of reaction.

The precise mechanism for these SmI_2-mediated couplings is not fully known. The reductive coupling could follow one of three possible pathways depending on the character of the carbonyl functionality (*e.g.* aldehyde, ketone, conjugated carbonyl) undergoing reductive carbon–carbon bond formation and whether the reaction is intra- or intermolecular. Three mechanistic scenarios can be envisaged (Scheme 5.2). Common for all three mechanisms is the initial reduction step involving a rapid and reversible inner-sphere electron transfer from Sm(II) to the C=O bond, resulting in the formation of the Sm(III)-bound ketyl radical anion. If the carbonyl group is easily reduced, then a high concentration of the ketyl is generated, which ultimately leads to dimerisation, possibly *via* a pseudo-bridged metal atom (path B). This mechanism is believed to operate for the intermolecular coupling of aldehydes and activated ketones.

In path A, it is possible that a second electron transfer from a second equivalent of SmI_2 reduces the ketyl, generating a metallaoxirane species. This reactive dianionic intermediate then adds to a second carbonyl substrate.

Organic Synthesis Using Samarium Diiodide: A Practical Guide
By David J. Procter, Robert A. Flowers, II and Troels Skrydstrup

Published by the Royal Society of Chemistry, www.rsc.org

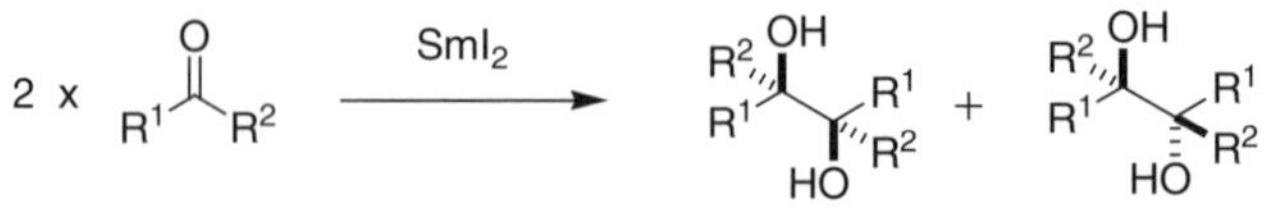

Scheme 5.1

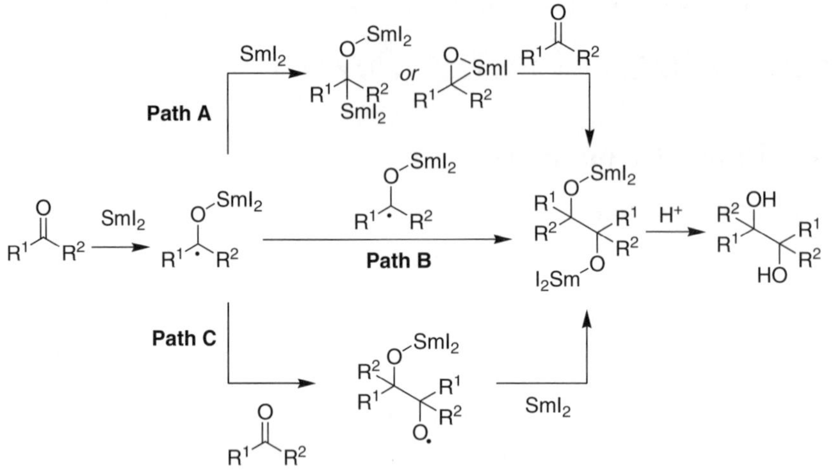

Scheme 5.2

Alternatively, the ketyl radical anion undergoes radical addition to the second carbonyl group, generating a highly reactive oxy radical species (path C). Although this addition is typically unfavourable, binding of the carbonyl substrate to the Lewis acidic Sm(III) centre of the ketyl radical anion would lead to a substantial LUMO activation. In addition, the highly reactive oxy radical intermediate is most likely reduced by SmI_2 immediately, resulting in the formation of a di-Sm(III) or a Sm(III)–diolate complex. This latter pathway can account for the observation that α-heteroatom-substituted carbonyl compounds undergo intramolecular pinacol couplings without cleavage of the α-heteroatom substituent, indicating that pinacol cyclisation is faster than the reduction of the ketyl radical anion to the corresponding anion. Furthermore, coordination of the metal centre to the second carbonyl group after the first electron-transfer step provides a nice explanation for the tendency for selective *cis*-diol formation in five- and six-membered ring syntheses.

Finally, there have been several exciting reports in recent years on the use of imine-type derivatives in pinacol coupling reactions for the stereoselective and asymmetric synthesis of vicinal amino alcohols and diamines. Although imines are generally less prone to undergo reduction to the ketyl radical anion with SmI_2, the imine nitrogen can be suitably functionalised with activating groups that facilitate reduction. The nitrogen also serves as an attachment point for a chiral auxiliary. This strategy has been used to great effect and will be discussed in Section 5.1.4.

5.1.2 Intermolecular Pinacol Couplings

SmI_2 can effectively promote the intermolecular reductive dimerisation of aldehydes or ketones giving rise to symmetrical diols.[7–9] Generally, arylaldehydes and aryl ketones couple within seconds in THF at room temperature. Aliphatic aldehydes and ketones react considerably more slowly: several hours are required for the aldehydes, whereas for ketones reaction times of 24 h are usually needed. Nevertheless, these slower couplings can be greatly accelerated by the addition of additives such as HMPA.[10]

Representative procedure (with tetraglyme). To a stirred solution of SmI_2 (1.3 equiv) in anhydrous degassed THF was added tetraglyme (1.3 equiv) under nitrogen and the reaction stirred for 10 min. Aldehyde (1 equiv) was added and the reaction mixture stirred for 5 min. The reaction was then quenched with aqueous saturated NH_4Cl and diluted with CH_2Cl_2. The organic phase was washed with H_2O ($\times 2$), dried (Na_2SO_4) and concentrated *in vacuo*. The crude product was purified by column chromatography (pentane–EtOAc eluent).

Representative procedure (without tetraglyme). Aldehyde (1 equiv) was dissolved in THF and added under nitrogen to a stirred solution of SmI_2 (1 equiv) in anhydrous degassed THF at room temperature. The reaction was then quenched with 0.1 M HCl followed by extraction with Et_2O. The organic phase was washed with aqueous saturated $Na_2S_2O_3$ and aqueous saturated NaCl, dried (Na_2SO_4) and concentrated *in vacuo*. The crude product was purified using column chromatography (pentane EtOAc eluent).

In most cases, particularly with simple aldehydes, both the *threo* and *erythro* stereoisomers are formed, with low stereocontrol. Only in the case of sterically hindered substrates such as 2,2-dimethylpropanal can high stereoselectivity be achieved in favour of the *threo* isomer (Scheme 5.3).[11] Some efforts have been made to increase the levels of stereoselectivity in these SmI_2-promoted pinacol couplings. Skrydstrup reported that the use of tetraglyme and derivatives thereof can provide a considerable increase in diastereoselectivity while maintaining good coupling yields (Scheme 5.3).[11] For 2-methylbutanal, the use of SmI_2 with tetraglyme results in improved selectivity for the *threo* product whereas the use of similar conditions in the pinacol coupling of benzaldehyde provided the *erythro* product preferentially.[12]

Uemura demonstrated that enantiomerically pure (arylaldehyde)tricarbonylchromium complexes afford exclusively the *threo*-pinacols, providing an asymmetric synthesis of hydrobenzoins (Scheme 5.4).[13] An intermediate involving coordination of the Sm(III) metal centre with the carbonyl oxygen was proposed to account for the high selectivity observed. This was supported by coupling experiments in the presence of HMPA, an additive that is known to prevent complexation, which led to the preferential formation of the *erythro* product.

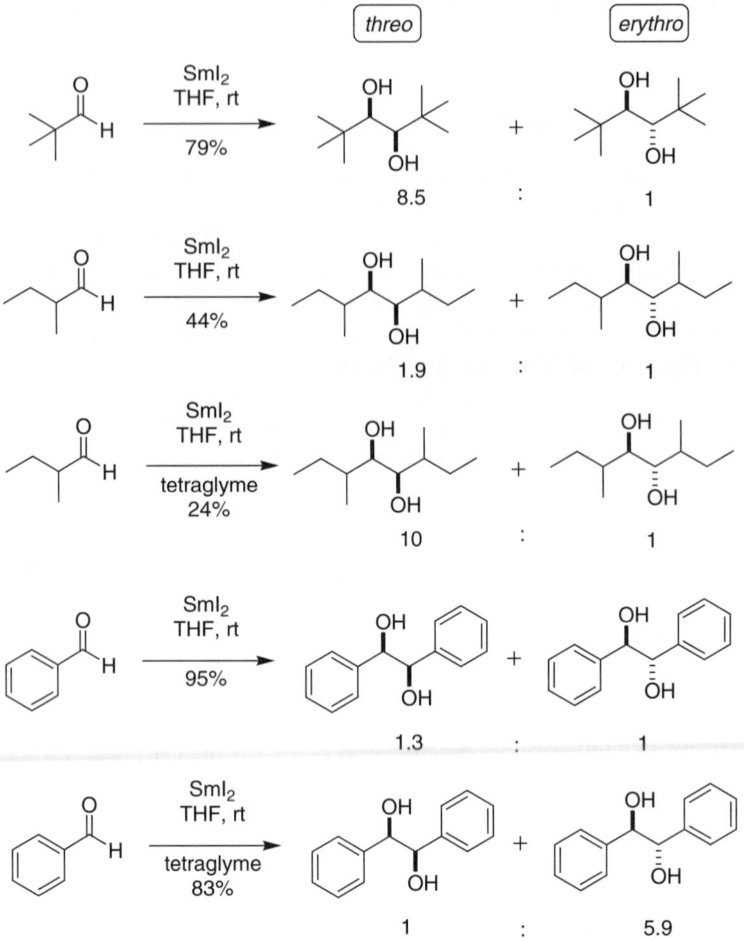

Scheme 5.3

In 1996, Endo reported a protocol for pinacol coupling using catalytic amounts of SmI_2 with magnesium as the stoichiometric reductant and TMSCl to liberate Sm(III) from the pinacolate (see Chapter 7, Section 7.2). The method provides good yields of the vicinal diol products and employs only 10 mol% of SmI_2 (Scheme 5.5).[14] Aspinall and Greeves later modified this protocol by using Me_2SiCl_2 as the chlorosilane and tetraglyme as an additive to improve the diastereoselectivity of the reductive couplings.[15]

5.1.3 Intramolecular Pinacol Couplings

The stereoselectivity of SmI_2-mediated pinacol couplings depends highly on the substrates used. In contrast to the intermolecular cases, high levels of diastereoselectivity can be attained in the cyclisation of a variety of 1,5- and

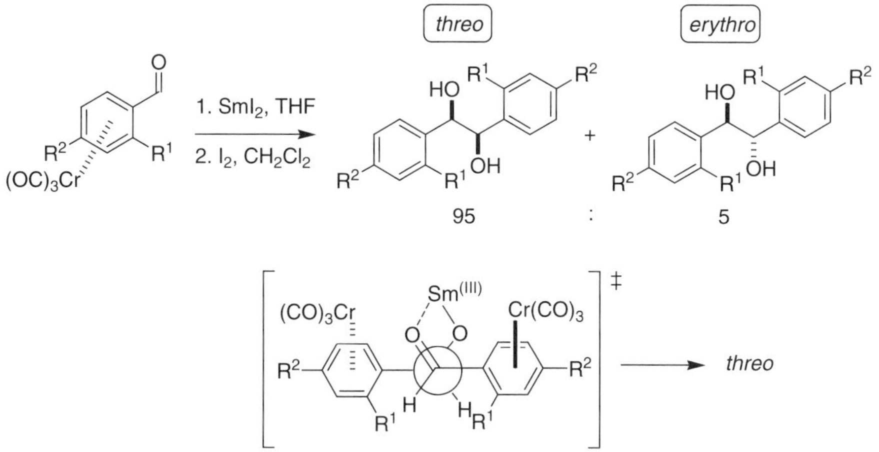

Scheme 5.4

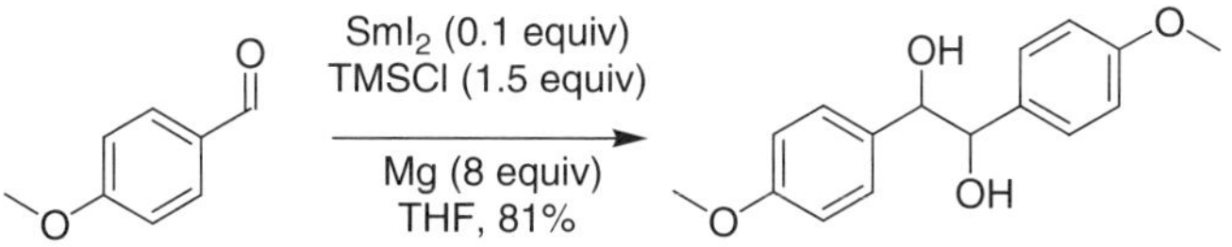

Scheme 5.5

1,6-dicarbonyl substrates, leading almost exclusively to the formation of *cis*-vicinal diols. Polar substituents α to the reacting carbonyl can act as controlling elements and generally end up *anti* to the newly formed *cis*-1,2-diols in the product. For example, in the synthesis of the carbocyclic monosaccharide caryose, Iadonisi reported an intramolecular SmI_2-promoted pinacol coupling of the ketoaldehyde **1** leading to the *cis*-diol **2** with a high degree of diastereocontrol (Scheme 5.6).[16]

Examples of chelation control from a neighbouring Lewis basic carbonyl have also been reported in the formation of five-membered cyclic diols. As illustrated in the cyclisation of the keto ester **3**, high diastereocontrol is observed for these SmI_2-mediated cyclisations (Scheme 5.7).[17,18] The asymmetric preparation of cyclisation substrates such as **4**, using Evans' chiral oxazolidinone auxiliary, allows access to highly functionalised and enantiomerically pure cyclopentane derivatives (Scheme 5.7). Unfortunately, the application of this type of chelation control for six-membered ring generation is less rewarding and low cyclisation yields and diastereoselectivities are often obtained.

Representative procedure. SmI_2 (2 equiv) in anhydrous degassed THF was cooled to −78 °C and a solution of carbonyl compound **3** (1 equiv) and MeOH (2 equiv) in dry THF was added dropwise under nitrogen. The resulting solution was stirred at

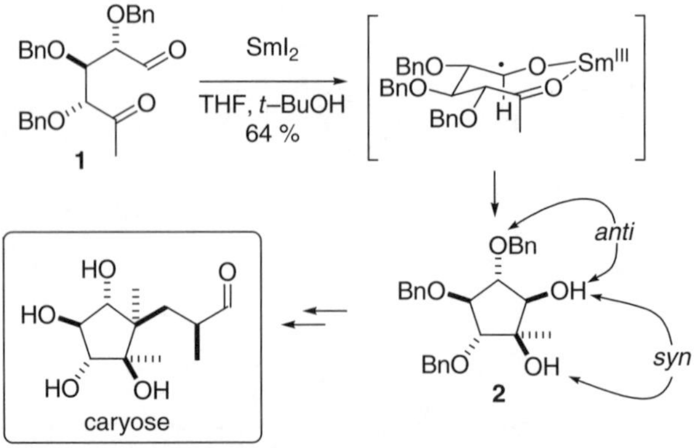

Scheme 5.6

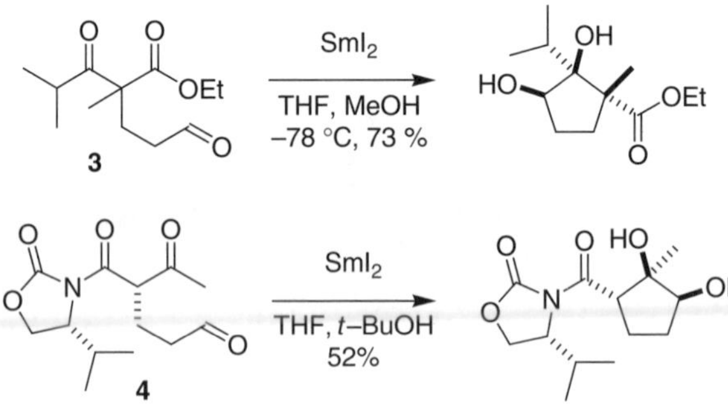

Scheme 5.7

−78 °C for 2 h and slowly warmed to room temperature. The reaction was then quenched with aqueous saturated $NaHCO_3$ and extracted into EtOAc. The organic phase was washed with aqueous saturated NaCl, dried (Na_2SO_4) and concentrated *in vacuo*. The crude product was purified by column chromatography (EtOAc–hexanes eluent).

Intramolecular pinacol coupling can also be used to prepare bicyclic ring systems, as demonstrated by Arseniyadis in a synthetic approach to paclitaxel (Taxol) (Scheme 5.8).[19] Treatment of ketoaldehyde **5** with SmI_2 furnished diol **6**, a precursor to the A ring of the target, in high yield.

There are many examples of six-membered ring formation using SmI_2-mediated pinacol couplings. Fernández-Mayoralas reported an interesting example of a stereoselective pinacol ring closure in his work on the synthesis of cyclitols as precursors for *C*-fucopyranosides.[20] The stereochemical outcome of the reaction triggered by treatment of the C_2-symmetrical dialdehyde **7** with SmI_2 was found to be highly dependent on the protecting groups on the hydroxyl substituents adjacent to the carbonyl functionalities (Scheme 5.9).

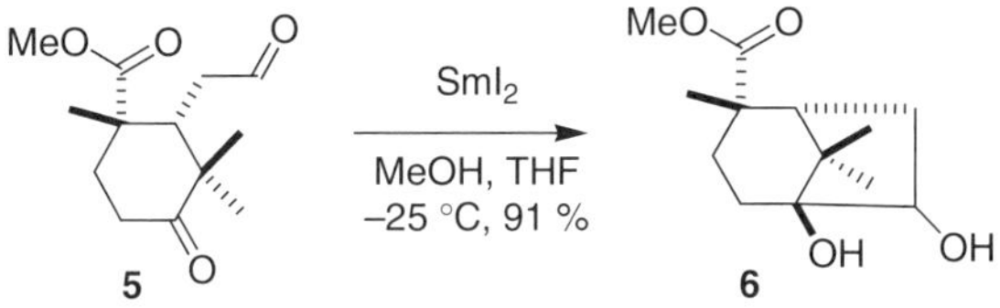

Scheme 5.8

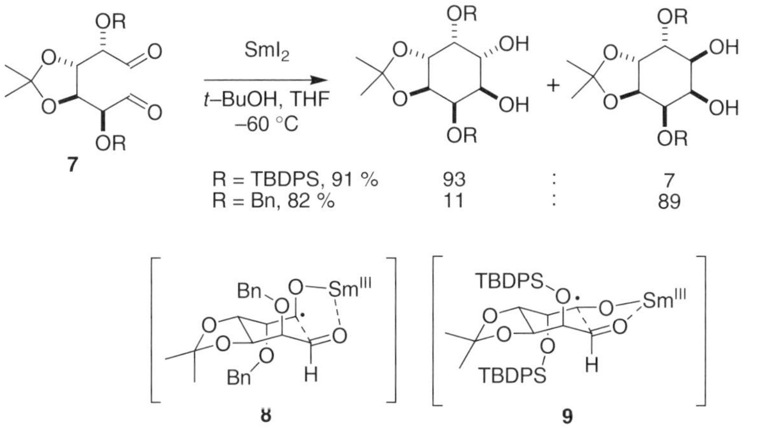

Scheme 5.9

The usual *cis*-diol product was preferentially obtained when benzyl protecting groups were present. This can be explained by invoking a transition structure **8** with a bridging Sm(III) ion between the ketyl radical anion and the aldehyde. In contrast, the use of TBDPS protecting groups led to selective generation of the *trans*-diol carbocycle: as the two α-substituents occupy pseudoaxial orientations during the cyclisation event, the use of the bulky silyl protecting group disfavours a transition structure analogous to **8** through adverse 1,3-diaxial interactions. Instead, an alternative transition structure **9**, free of these unfavourable interactions, leads to the formation of the *trans*-diol.

Another example of the impact of neighbouring protected hydroxyl groups on the diastereoselectivity of pinacol couplings can be found in d'Alarcao's studies on the synthesis of the *myo*-inositol derivative **10** (Scheme 5.10).[21]

With the non-C_2-symmetrical dialdehyde **11**, reductive cyclisation on treatment with SmI_2 led to the preferential formation of the cyclic *cis*-diol **12**. The intervention of the transition state structure **13** was invoked to rationalise the *anti*-selectivity observed with the bulky silyl ether-protected hydroxyl group.

The intramolecular SmI_2-mediated pinacol coupling has been well exploited as a key step in many natural product syntheses for the construction of different ring sizes. For example, in 1997, Pancrazi reported a formal total synthesis of

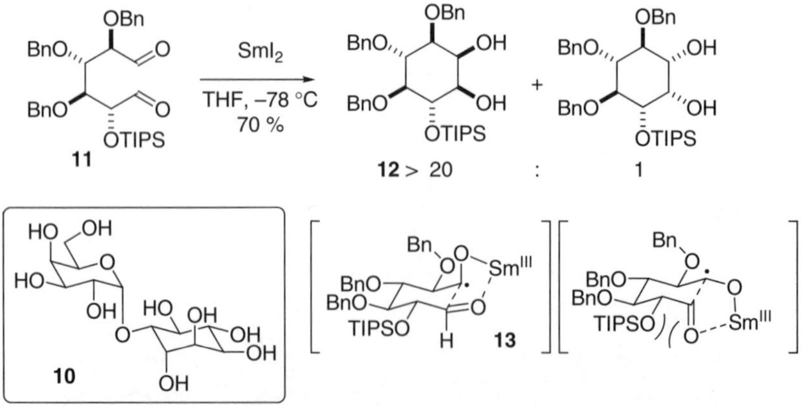

Scheme 5.10

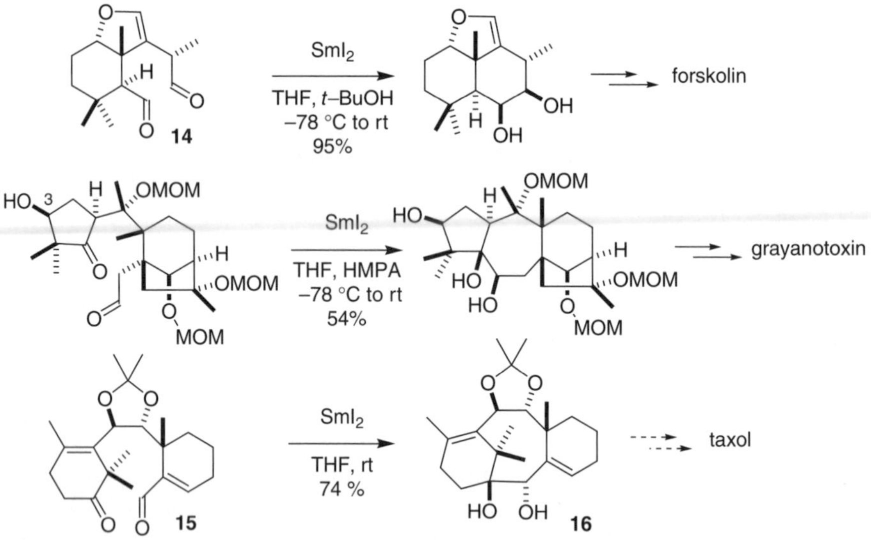

Scheme 5.11

forskolin, in which the central six-membered ring was efficiently created by cyclisation of the dialdehyde **14** (Scheme 5.11).[22]

In Matsuda's total synthesis of grayanotoxin, a combination of SmI_2 and HMPA was used to secure the seven-membered carbocycle with complete diastereocontrol at the two new stereogenic centres (Scheme 5.11).[23] It is interesting that the unprotected C3-hydroxyl group was essential for achieving ring closure, as the corresponding C3-MOM ether derivative did not undergo cyclisation. Finally, an example of eight-membered ring formation using a SmI_2-promoted pinacol coupling was provided by Swindell in his approach to the B-ring of the taxane skeleton:[24] treatment of ketoaldehyde **15** with SmI_2 gave *anti*-diol **16** in 74% yield (Scheme 5.11). The value of SmI_2-mediated cyclisations in natural product synthesis is highlighted in Chapter 7, Section 7.4.

5.1.4 Pinacol Couplings of Imines and Their Equivalents

Pinacol couplings in which one of the coupling partners is an imine derivative are particularly useful for the construction of vicinal amino alcohols. In particular, oximes and hydrazones have proven their worth in heteropinacol cyclisation reactions and have been used in the construction of five- to seven-membered carbo- and heterocycles. Particularly noteworthy is the observation that *trans* products are often the major products formed from these cyclisations. Thus the diastereoselectivity of heteropinacol couplings often contrasts with analogous pinacol cyclisations of dicarbonyl compounds, discussed above in Section 5.1.3.[25,26] For example, in the synthesis of an adenosine analogue, the aldehyde-oxime **17** cyclised in good yield to give pyrrolidine **18** with a 9:1 preference for the *trans* isomer (Scheme 5.12).[27] In these cyclisations, the alkoxyamine product can be reduced directly to the corresponding amine by a second reduction event with SmI_2 either in a follow-up reaction or through the use of excess SmI_2 in the ring-forming step.[28] (For a discussion of N–O bond reduction using SmI_2, see Chapter 4, Section 4.6.2.)

Representative procedure. To a solution of oxime **17** (1 equiv) and *t*-BuOH (2.5 equiv) in THF was added dropwise SmI_2 (3 equiv) in anhydrous degassed THF under an argon atmosphere at –78 °C. The solvent was then evaporated at reduced pressure. The resulting residue was taken up in Et_2O and filtered through a pad of Celite. The filtrate was concentrated *in vacuo* and the crude product was purified by column chromatography (EtOAc–hexanes eluent).

As oximes and hydrazones are not good electron acceptors, the mechanism for these reactions most likely involves an intermediate ketyl radical anion, which adds to the C=N double bond. A three-electron–two-orbital interaction involving the nitrogen-centred radical with the lone pair of the adjacent

Boc–N 17 N–OBn O; SmI_2, THF, *t*-BuOH, –78 °C, 70 %; Boc–N, N(H)–OBn, OH **18** *major*, dr 9:1

HN, N, NH₂, OH — an adenosine analogue

Scheme 5.12

heteroatom explains the good radical acceptor abilities of these imine-type derivatives.[26]

An impressive application of an aldehyde–oxime coupling was reported by Nicolaou in their studies on the total synthesis of the complex natural product diazonamide A (see Chapter 7, Section 7.4).[29,30] Naito[31] and Skrydstrup[32] have reported impressive applications of the heteropinacol cyclisation for the stereoselective synthesis of the azepine ring of the PKC-inhibitor balanol (see Chapter 7, Section 7.4).

Analogous intermolecular couplings have only recently been shown to be a viable strategy for the preparation of vicinal amino alcohols. In 2002, Py and Vallée disclosed the remarkable ability of nitrones to undergo intermolecular heteropinacol couplings with aldehydes and ketones, generating *N*-hydroxyamino alcohols in good yields although with low diastereoselectivities (Scheme 5.13).[33] Cyclic nitrones are also worthy coupling partners for the preparation of α,α-disubstituted pyrrolidine methanols.[34]

Interestingly, cyclisation of the ketone-nitrone **19** and concurrent N–O bond reduction with excess SmI_2 provided exclusively the *cis*-amino alcohol, thereby complementing similar cyclisation reactions with oximes and hydrazones that typically give *trans* products (Scheme 5.14).[33] An alternative mechanism was

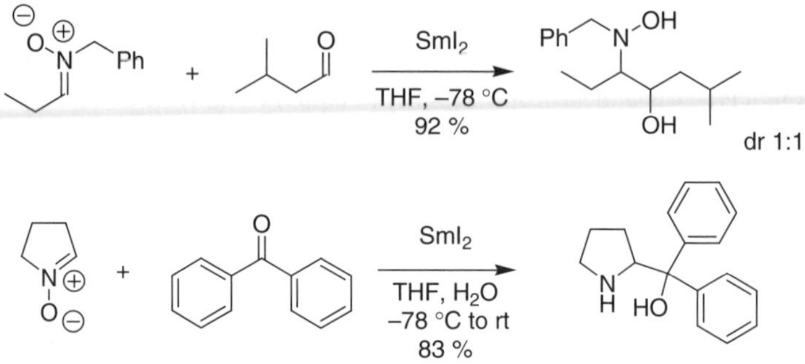

SmI_2, THF, –78 °C, 92 %; dr 1:1

SmI_2, THF, H_2O, –78 °C to rt, 83 %

Scheme 5.13

SmI_2, THF, –78 °C to rt, 75 %; **19**; *major* dr 7:1

Sm^{III}, R^1, R^2, R^3, R^4, R^5

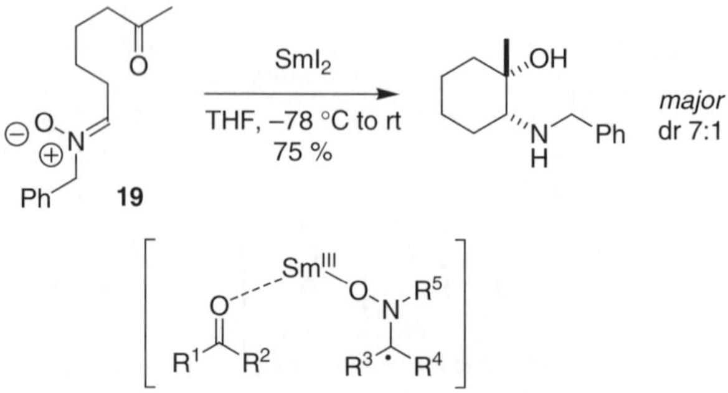

Scheme 5.14

proposed for these reductive coupling reactions, in which the nitrone is preferentially reduced to a ketyl radical anion-like intermediate followed by addition to the carbonyl group *via* a chelated transition structure.

Representative procedure. A stirred and carefully deoxygenated solution of nitrone **19** (1 equiv) in dry THF was cooled to –78 °C under argon. A solution of SmI_2 (2 equiv) in anhydrous degassed THF was added and the reaction maintained at –78 °C until conversion was complete. Aqueous saturated solutions of $Na_2S_2O_3$ and $NaHCO_3$ were added and the mixture was extracted with EtOAc. The organic layer was washed with aqueous saturated NaCl, dried (Na_2SO_4) and concentrated *in vacuo*. The crude product was purified by column chromatography (CH_2Cl_2–methanol eluent).

Prior to 2005, there were few examples of such stereoselective couplings; for example, Uemura had reported the cross-pinacol coupling of planar chiral ferrocenecarboxaldehyde with imines in 2002.[35] In 2005, Xu reported an exceptional solution to the challenge of carrying out stereoselective intermolecular heteropinacol couplings.[36] It was shown that the SmI_2-induced reductive coupling of a series of *N-tert*-butylsulfinylimines with alkylaldehydes provided the pinacol products in high yield and with excellent diastereoselectivity. Subsequent acidic cleavage of the sulfinyl group led to the corresponding β-amino alcohols in high enantiomeric excess. Xu exploited this approach for the asymmetric synthesis of a variety of bioactive compounds such as the human neurokinin-1 substance P receptor antagonist (+)-L-733,060 (Scheme 5.15).

SmI2 t–BuOH THF, –78 °C 89 %

HCl MeOH

major dr 98:2, >99% ee

SmI2 t–BuOH THF, –78 °C 78 %

11 steps

Scheme 5.15

Representative procedure. SmI_2 (1.3 equiv) in anhydrous degassed THF was cooled to –78 °C under nitrogen and a solution of *t*-BuOH (1.3 equiv) and aldehyde (1 equiv) in dry THF was added dropwise. *N-tert*-Butylsulfinylimine (0.7 equiv) in dry THF was then added dropwise. The reaction was quenched with aqueous saturated $Na_2S_2O_3$ and extracted into EtOAc, then dried (Na_2SO_4) and concentrated *in vacuo*. The crude product was purified by column chromatography.

Both simple alkyl- and arylaldimines can undergo SmI_2-mediated reductive dimerisation; however, the diastereoselectivity of these reactions is generally of limited use.[37,38] In recent years, Xu has published a series of exciting papers describing the asymmetric synthesis of vicinal diamines *via* homo- and cross-pinacol coupling of imine derivatives. SmI_2-mediated reductive homocoupling of aromatic *N-tert*-butylsulfinylimines proved to be effective upon treatment with SmI_2–HMPA, providing the C_2-symmetrical vicinal diamines as single stereoisomers (Scheme 5.16).[39] A mechanistic proposal was put forth involving a ketyl-like intermediate **20**, in which the *s-trans* conformer is the most stable due to the bulkiness of HMPA-bound Sm(III). The chiral auxiliary then directs *Re*-face approach of the two ketyl-like radical anions leading to the homo-coupled product.

Representative procedure. SmI_2 (2 equiv) in anhydrous degassed THF was cooled to –78 °C under nitrogen and a solution of HMPA (2 equiv) in dry THF was added dropwise. After 30 min, sulfinylimine (1 equiv) in dry THF was added dropwise. The reaction was quenched with aqueous saturated $Na_2S_2O_3$, then extracted into EtOAc and dried (Na_2SO_4) and concentrated *in vacuo*. The crude product was purified by column chromatography.

More impressive still are the cross-coupling results reported by the same group for the asymmetric synthesis of unsymmetrical diamines. In the absence of HMPA, SmI_2 promoted the cross-coupling of nitrones with *N-tert*-butyl-sulfinylimines again with high stereocontrol at the two new stereogenic centres

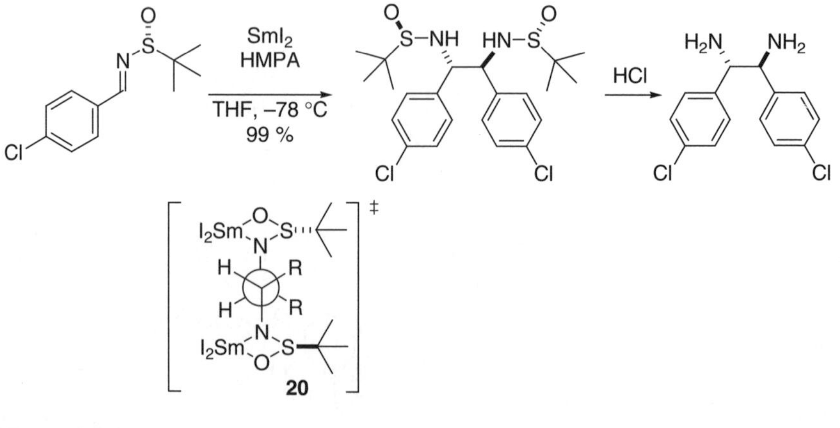

Scheme 5.16

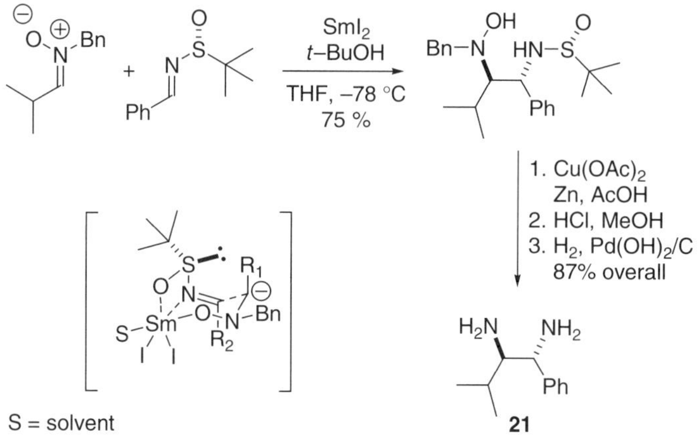

Scheme 5.17

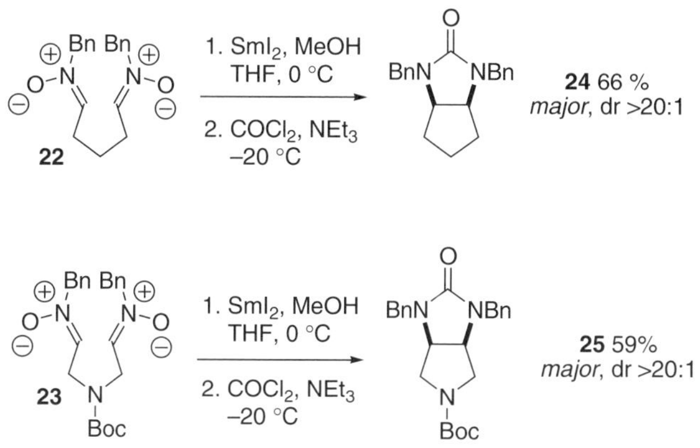

Scheme 5.18

(Scheme 5.17).[40] A three-step reduction–deprotection protocol liberated the *anti*-vicinal diamine **21**. A six-membered cyclic transition-state structure was proposed to account for the *anti* selectivity after a two-electron reduction of the nitrone. The chiral *N-tert*-butylsulfinyl group directs the attack of the carbanion to the *Si*-face of the C=N double bond of the imine.

Finally, Skrydstrup reported that five-membered ring systems containing a *cis*-diamine motif can be prepared using a SmI_2-mediated reductive cyclisation of dinitrones.[41] For example, treatment of dinitrones **22** and **23** with excess SmI_2 resulted in cyclisation followed by N–O bond cleavage. Trapping of the ensuing diamine with phosgene produced the cyclic ureas **24** and **25** in good overall yield (Scheme 5.18). In all cases, *cis*-diamines were the major products obtained, hence the coupling shows similarities to the intramolecular pinacol couplings between two carbonyl groups. Surprisingly, attempts to promote the cyclisation of nitrones with hydrazones or oximes, as previously reported for aldehydes and ketones, were not successful.

Representative procedure. Dinitrone **22** (1 equiv) was dissolved in dry THF and dry MeOH (16 equiv) was added. The reaction was cooled to 0 °C and SmI_2 (8 equiv) in anhydrous degassed THF was added dropwise over 30 min. Upon completion, the reaction was quenched with aqueous saturated $Na_2S_2O_3$ and aqueous $NaHCO_3$ was then added. The mixture was extracted into EtOAc and the combined organic layers were washed with aqueous saturated NaCl, dried (Na_2SO_4) and concentrated *in vacuo* to yield 1,2-diamine. The crude diamine was dissolved in CH_2Cl_2 and Et_3N (2.2 equiv) and cooled to –20 °C. A solution of phosgene (20% in toluene, 1 equiv) was added dropwise and the reaction allowed to warm to room temperature. Upon completion, aqueous NaOH (5 M) was added and the mixture extracted into CH_2Cl_2, dried (Na_2SO_4) and concentrated *in vacuo*. The crude product was purified by column chromatography (pentane–EtOAc eluent).

5.1.5 Conclusions

SmI_2-promoted pinacol couplings have evolved significantly over the years to become one of the most useful reactions for the stereoselective preparation of 1,2-diols in acyclic and cyclic systems. The recent extension of these SmI_2-mediated reactions to the asymmetric synthesis of β-amino alcohols and vicinal diamines through the coupling of a variety of imine derivatives represents a significant advance. The discovery of powerful, new SmI_2-mediated pinacol coupling procedures will ensure that the transformation continues to enjoy widespread application in target synthesis.

5.2 Carbonyl–Alkene Couplings

5.2.1 Introduction

Carbonyl–alkene couplings represent one of the most widely studied and most useful carbon–carbon bond-forming reactions promoted by SmI_2. An extensive array of coupling partners has been investigated in a remarkable range of intramolecular and intermolecular reactions. Fundamentally, the reaction can be considered as a highly effective method for the reductive coupling of aldehydes or ketones to alkenes, alkynes and aromatic systems, with moderate to high diastereocontrol. Furthermore, recent developments include the use of a variety of reactive imine equivalents such as nitrones and sulfinimines, in place of the aldehyde or ketone component (see Section 5.1), which provides an alternative means of accessing chiral amines. As with many other SmI_2-mediated transformations, fine-tuning of the reaction using additives such as proton sources or HMPA can have a marked influence on the efficiency and the stereochemical outcome of the reaction.

As expected for intermolecular radical addition reactions involving a nucleophilic radical donor, electron-deficient alkenes such as acrylates, acrylamides and acrylonitriles represent the best coupling partners in intermolecular SmI_2-promoted carbonyl–alkene couplings. Mechanistically, these reactions have been described as ketyl–alkene couplings, involving single-electron reduction of the carbonyl substrate to give a ketyl radical anion intermediate followed by addition to the unsaturated carbon–carbon bond. Subsequent reduction of the new radical intermediate leads to a Sm(III) enolate or enolate-type species, which abstracts a proton from an acidic source (Scheme 5.19). Although this reaction mechanism has been accepted for many years, more recent evidence has shown that an alternative mechanism may also take place when electron-deficient alkenes are involved. This alternative

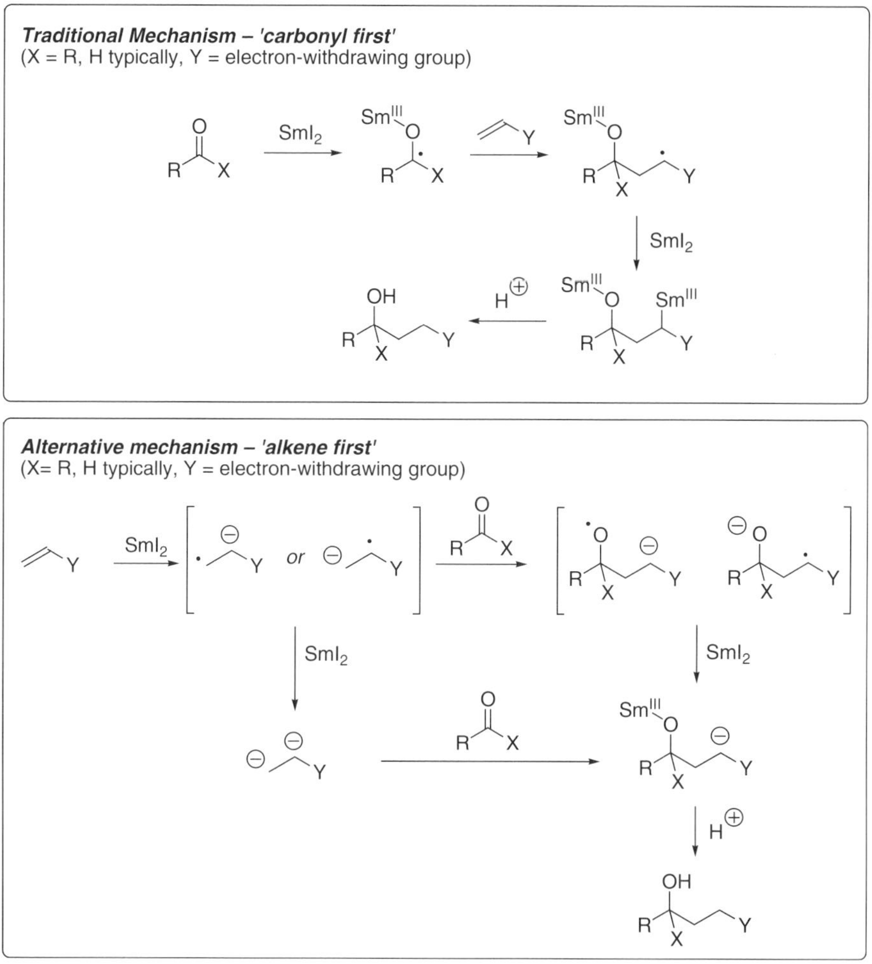

Traditional Mechanism – 'carbonyl first'
(X = R, H typically, Y = electron-withdrawing group)

SmI_2 / Sm^{III} / Y / SmI_2 / Sm^{III} / $H^{\oplus}$ / OH / R / X / Y

Alternative mechanism – 'alkene first'
(X= R, H typically, Y = electron-withdrawing group)

SmI_2 / or / SmI_2 / SmI_2 / Sm^{III} / $H^{\oplus}$ / OH / R / X / Y

Scheme 5.19

mechanism involves selective conjugate reduction of the double bond followed by anionic (or possibly radical) addition to the carbonyl group (Scheme 5.19). Therefore, the precise mechanism for a given reductive carbonyl–alkene coupling will depend on the specific combination of reactants and the respective rates of reduction of each component by the lanthanide reagent.

Procter has suggested that a study of the dependence of reaction outcome on the stereochemistry of the alkene in the substrate can be used to gain information on the mechanistic 'direction' of reductive couplings.[42,43] In cases where the alkene stereochemistry has a marked effect on the reaction outcome, a traditional 'carbonyl first' mechanism may be in operation, whereas in reactions where alkene stereochemistry has little effect, an alternative mechanism in which the alkene is reduced first and a common reactive intermediate is formed, regardless of the geometry of the starting alkene, may operate (Scheme 5.20).[42,43]

Additional support for an alternative 'alkene first' mechanism in these cyclisations came from the successful cyclisation of cyclopropyl ketone **26** with no observance of fragmentation products (Scheme 5.21).[43] This suggests that a ketyl radical anion is not formed during the cyclisation and that reduction of the alkene leads to product formation.

*Intra*molecular carbonyl–alkene coupling reactions often are efficient and display a greater degree of tolerance in the alkene component than the *inter*molecular variant. Most notably, unactivated alkenes are willing participants in cyclisation reactions. In these cases, the reactions must proceed through the 'traditional' pathway involving ketyl radical anions as the alkene cannot be

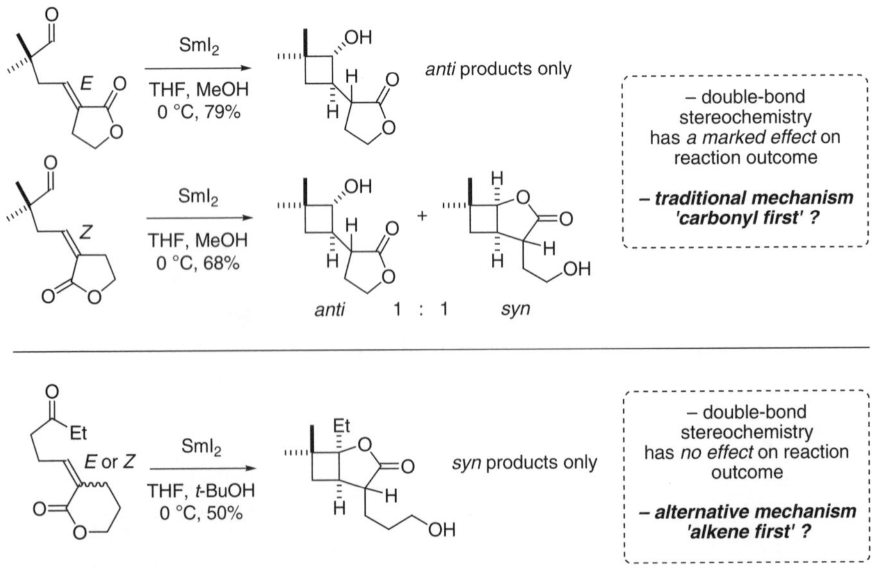

Scheme 5.20

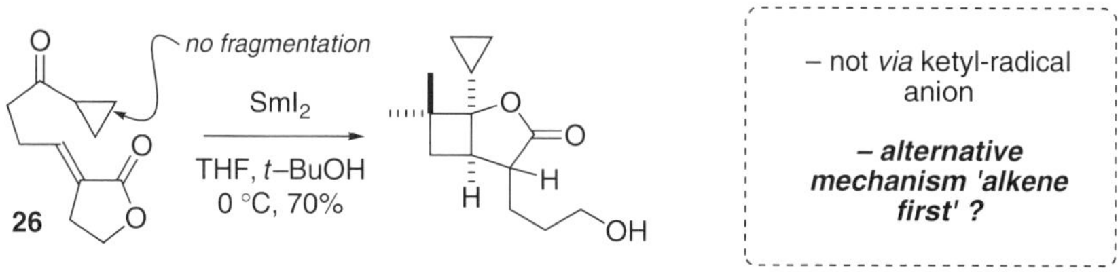

Scheme 5.21

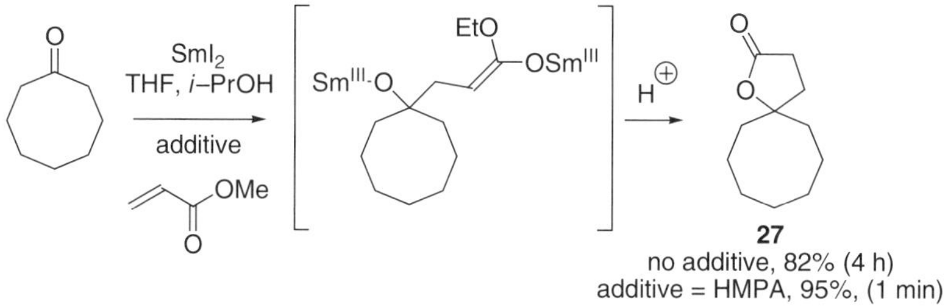

Scheme 5.22

reduced under the reaction conditions. A large variety of cyclic alcohols possessing small- to medium-sized rings can be assembled under mild conditions and generally with a high degree of stereochemical control using the reaction.

5.2.2 Intermolecular Carbonyl–Alkene Couplings

Fukuzawa[44,45] and Inanaga[46] showed that the coupling of ketones and aldehydes with acrylates mediated by SmI_2 generally leads to γ-lactones due to cyclisation of the intermediate alkoxide produced from the carbon–carbon bond-forming step. This is illustrated by the reaction of cyclooctanone with methyl acrylate to provide the spirolactone **27** in high yield (Scheme 5.22).

Representative procedure. To a stirred solution of SmI_2 in THF (3 equiv) at 0 °C was added a solution of ketone (1 equiv), α,β-unsaturated ester (2 equiv) and 2-propanol (1.5 equiv) in THF dropwise and the reaction mixture was stirred for between 3 and 6 h. The reaction was quenched by opening to air and the addition of aqueous, saturated $Na_2S_2O_3$, followed by extraction with Et_2O. The organic extracts were combined, dried (Na_2SO_4) and the solvent removed *in vacuo*. The crude product was purified by column chromatography (EtOAc–hexane eluent). [Optionally, HMPA (1 ml per 1 mmol of starting material) was added prior to addition of the reactants and the colour of the solution changed from blue–green to purple–black. The reaction mixture was stirred for 1 min after addition of the reactants.]

Essentially all carbonyl–alkene coupling reactions are carried out in the presence of a proton source, typically an alcohol. The use of a protic additive is essential for obtaining high yields of the coupled product. Although the additive can perform different functions in these SmI_2-mediated reactions (see Chapter 2, Section 2.2), one important role of the alcohol additive is undoubtedly to neutralise basic intermediates generated after the radical addition step, for example, the Sm(III) enolate in Scheme 5.22. The addition of HMPA can also have a positive effect, accelerating substantially the coupling reaction by the generation of a more powerful reducing agent upon complexation of the Lewis basic additive to SmI_2.[46] The inclusion of a chelating substituent α to the carbonyl group of the Sm(III) ketyl precursor has been shown to induce high levels of asymmetric induction in the formation of adducts. For example, Matsuda has shown that treatment of α-hydroxy ketone **28** with SmI_2 and MeOH as the proton source resulted in reductive addition of **28** to enoates, including sterically hindered β-substituted enoates, giving γ-lactones with high levels of diastereocontrol at the two contiguous stereogenic centres (Scheme 5.23).[47] The formation of a five-membered ring chelate after the first electron transfer step explains the stereoselectivity observed in these reactions. As might be expected, the addition of HMPA reduces the diastereoselectivity of these reactions, by preventing the formation of the chelated ketyl radical anion.

In 1997, Fukuzawa reported an asymmetric synthesis of γ-lactones through the reductive coupling of aldehydes and ketones with acrylates and crotonates bearing the chiral auxiliary *N*-methylephedrine.[48] The reaction of numerous aldehydes and ketones with either (1*R*,2*S*)- or (1*S*,2*R*)-*N*-methylephedrinyl acrylate led to the formation of γ-lactones in high enantiomeric excess. Similarly, crotonates derived from *N*-methylephedrine provided *cis*-3,4-substituted γ-lactones as the major products with enantioselectivities ranging from 83 to 97%. For example, exposure of a mixture of cyclohexanecarboxaldehyde and *N*-methylephedrinyl crotonate **29** to SmI_2 and *t*-BuOH gave **30** in excellent yield with near perfect control of absolute and relative stereochemistry (Scheme 5.24).[48] Particularly noteworthy in Fukuzawa's approach is the liberation of the chiral auxiliary during the spontaneous lactonisation step, which can then be isolated and re-used in subsequent reductive couplings. This

Scheme 5.23

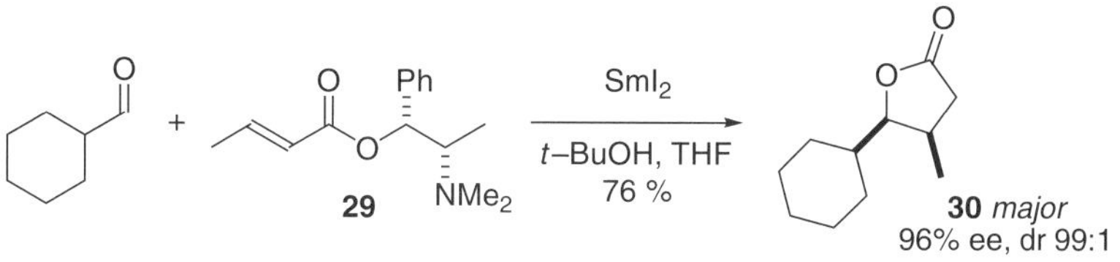

Scheme 5.24

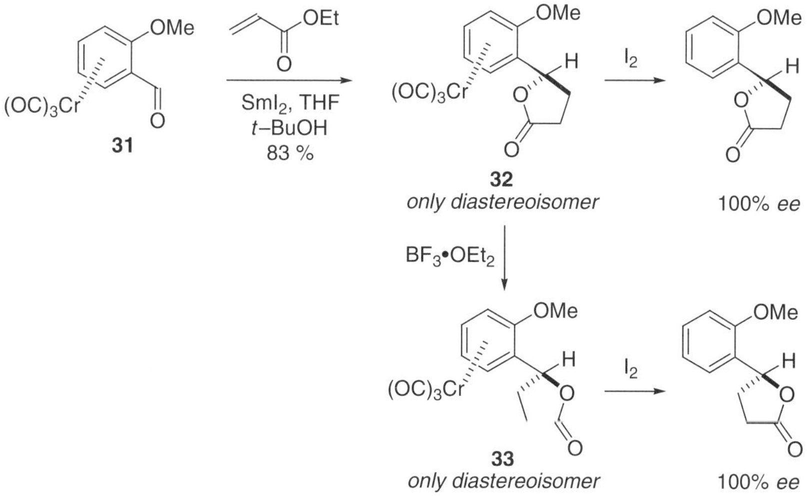

Scheme 5.25

reaction has been adapted by Procter in a solid-phase, asymmetric catch-and-release approach to γ-lactones using polymer-supported acrylates and crotonates and an ephedrine linker (see Chapter 7, Section 7.3).[49,50] Xu investigated the use of other chiral auxiliaries in the transformation for the preparation of enantiomerically enriched γ-lactones.[51]

> *Representative procedure.* To a stirred solution of SmI_2 in THF (2.2 equiv) at 0 °C was added slowly (0.33 ml min^{-1}) a solution of aldehyde (1 equiv), (1*R*,2*S*)-*N*-methylephedrinyl crotonate (1.2 equiv) and *tert*-butanol (1 equiv) in THF (1 M) and the reaction mixture was stirred at 0 °C for 3 h. The reaction was quenched by the addition of dilute HCl, followed by extraction with Et_2O. The organic phase was washed with aqueous, saturated $Na_2S_2O_3$ and brine, dried (Na_2SO_4) and the solvent removed *in vacuo*. The crude product was purified by column chromatography (Et_2O–hexane eluent).

Alternatively, enantiomerically enriched 3-aryl-γ-lactones have been synthesised using a SmI_2-mediated carbonyl–alkene coupling between enantiomerically pure arylaldehyde–$Cr(CO)_3$ complexes and ketones (Scheme 5.25).

For example, coupling of complexed aldehyde **31** with ethyl acrylate gave lactone **32** with complete diastereocontrol.[52] Furthermore, Lewis acid-mediated epimerisation of lactone **32** allowed for the isolation of the alternative diastereoisomer **33**. Subsequent decomplexation with iodine liberated the two enantiomerically pure lactones.[52]

5.2.3 Intramolecular Carbonyl–Alkene Couplings

SmI_2-mediated intramolecular carbonyl–alkene couplings are by far the most investigated variant of these carbon–carbon bond-forming reactions, allowing for the preparation of a wide range of carbo- and heterocyclic systems of varying ring size. As expected, these cyclisation reactions are more tolerant of the type of alkene than their intermolecular counterparts, thereby extending the reaction to unactivated alkenes. The use of SmI_2 to mediate such couplings has an additional benefit in that good levels of stereocontrol are typically observed in many of the cyclisations (Scheme 5.26). The SmI_2-mediated reductive cyclisation of isolated ketones on to unactivated alkenes represents the most straightforward of these intramolecular radical addition reactions. There are many examples of 5-*exo*-cyclisations and, in the simplest cases, stereoselectivities depend markedly on the substituent on the ketone as illustrated in Scheme 5.26.[53] In all circumstances, HMPA is required to allow the ketyl additions to run with useful reaction times. In nearly all cases, the outcome of 5-*exo*-carbonyl–alkene cyclisations can be explained by invoking a chair-like *anti* transition structure, in which the newly forming carbon radical centre is eclipsed by the ketone substituent.[53] Increasing the size of the substituent disfavours the *anti* transition structure and the diastereoselectivity can be reversed.

In general, high levels of diastereoselectivity are observed for 5-*exo*-cyclisations leading to both fused and bridged bicyclic ring systems. For example,

SmI_2
THF, HMPA
t–BuOH

R	Yield	Ratio
R = Me	86 %	> 150 : 1
R = *i*-Pr	85 %	23 : 1
R = *t*-Bu	78 %	3 : 1
R = Ph	48 %	< 1 : 150

$OSmI_2L_n$ — *anti*-transition structure **typically favoured**

$OSmI_2L_n$ — *syn*-transition structure **typically disfavoured**

Scheme 5.26

Scheme 5.27

Scheme 5.28

cyclisation of ketone **34** gave bicyclic alcohol **35** with complete diastereocontrol (Scheme 5.27). This high level of control can be explained by the imposing constraint of the additional ring.[53,54]

Representative procedure. To a stirred solution of SmI_2 in THF (2.8 equiv) was added HMPA (20 equiv) and the colour of the solution changed from blue–green to purple–black. After stirring the solution for 10 min, the unsaturated ketone (1 equiv) and *tert*-butanol (3 equiv) in THF (0.05 M) were added slowly (1 ml min^{-1}) and the mixture was stirred until reaction was complete. The reaction was quenched by opening to air and the addition of aqueous, saturated $NaHCO_3$, followed by extraction with Et_2O. The organic extracts were combined, washed again with H_2O and brine, dried (Na_2SO_4) and the solvent removed *in vacuo*. HMPA was removed by short column chromatography and the crude product was purified by column chromatography (EtOAc–hexane eluent) or Kugelrohr distillation.

Alkynes can also be used as radical acceptors for the generation of cyclopentanols bearing an exocyclic alkene, but in nearly all cases the yields are modest compared with those obtained in the related cyclisations of alkenes. Activation of the alkynes by electron-withdrawing groups or silyl substituents leads to more efficient 5-*exo-dig* ring closure.[55]

In 2007, Molander demonstrated that 5-*exo*-transannular reductive carbonyl addition to an endo- or exocyclic alkene can be highly effective for the formation of bicyclic ring systems. For example, treatment of cyclooctenone **36** with SmI_2 and HMPA gave bicyclic alcohol **37** in 70% yield (Scheme 5.28).[56] Transannular cyclisations have even been extended to 11-membered carbocyclic starting materials and give bicyclic products in good yield.

The presence of a suitably placed substituent that is capable of chelation to Sm(II) and Sm(III) in substrates for 5-*exo*-radical cyclisation has two advantages: first, HMPA is not required to promote the initial reduction step as the electron-withdrawing nature of the substituent lowers the LUMO of the ketone carbonyl, thereby facilitating the electron transfer into the group, and second, chelation of the Sm(III) in the ketyl intermediate to the proximal group leads to products with high diastereocontrol. This is best illustrated when a Lewis basic carbonyl group is present α to the ketone carbonyl as in **38**. Treatment of **38** with SmI_2 gave **39** as a single diastereoisomer in 78% yield (Scheme 5.29).[18]

In addition to the popular five-membered ring-forming reactions with non-activated alkenes, a few publications have revealed the efficacy of the 6-*exo* variant for generating cyclohexanols, and also more elaborate bicyclic alcohols.[53,54] Here too the yields for ring closure are high and comparable to those obtained in the syntheses of cyclopentanols. Even more impressive are Molander's results on 8-*endo*-cyclisations: treatment of unsaturated ketone **40** with SmI_2 gave cyclooctanol **41** in a synthetically useful yield (Scheme 5.30).[57] The substitution pattern, however, does have a marked effect on the efficiency of these reactions. The ability of SmI_2 to promote the formation of medium-sized rings, in addition to smaller cycles, can be attributed to the high steric bulk of the alkyoxysamarium substituent with as many as six HMPA molecules coordinated to the metal centre. The ketyl radical anion is thereby protected from alternative reaction pathways such as hydrogen abstraction from the ethereal solvent.

Molander exploited these remarkable 8-*endo* cyclisations in an asymmetric total synthesis of (+)-isoschizandrin (Chapter 7, Section 7.4).[58]

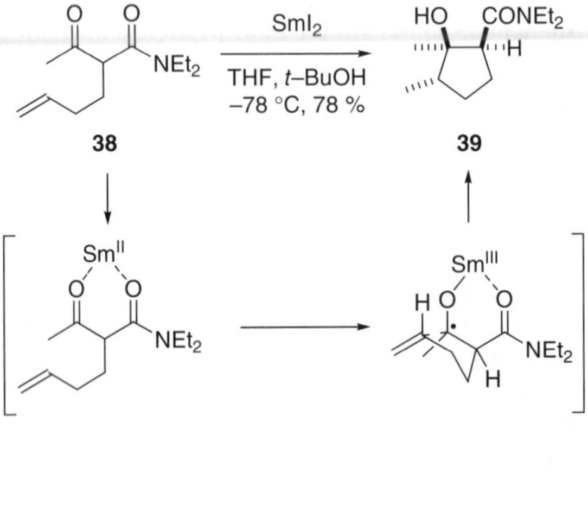

Scheme 5.29

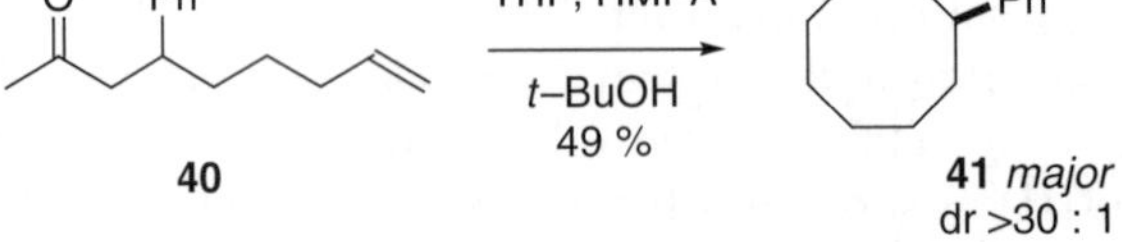

Scheme 5.30

The use of activated alkenes in these ring-forming events has been extensively studied and a variety of cyclic systems have been prepared. In many of these cases, the use of HMPA cosolvent is not necessary for achieving high conversions. The omission of HMPA is particularly important where chelation control is required to deliver products with high diastereoselectivity. For example, Weinges[59] and Procter[60,61] have reported an efficient 4-*exo*-cyclisation of γ,β-unsaturated aldehydes to give *anti*-cyclobutanols in good yield (Schemes 5.20 and 5.31). The cyclisation was employed by Procter in an approach to the natural product pestalotiopsin A (Chapter 7, Section 7.4).

Representative procedure. To a stirred solution of SmI_2 in THF (2 equiv) at 0 °C was added MeOH (25% by volume of the THF used in reaction). After stirring for 10 min, a solution of unsaturated aldehyde (1 equiv) in THF (0.3 M) was added and the reaction mixture was stirred for 5 min at 0 °C. Aqueous saturated NaCl and citric acid (2 equiv) were added and the reaction mixture was allowed to warm to room temperature. The aqueous layer was separated and extracted with EtOAc. The combined organic extracts were dried (Na_2SO_4) and concentrated *in vacuo*. The crude product was purified by column chromatography (EtOAc–hexane as eluent).

SmI_2-mediated carbonyl–alkene cyclisations are also effective when the alkene contains a β-substituent. For example, Enholm reported the SmI_2-mediated cyclisation of aldehyde **42** in the presence of MeOH to provide the highly functionalised cyclopentanol **43** containing a quaternary stereocentre, in good yield (Scheme 5.32).[62]

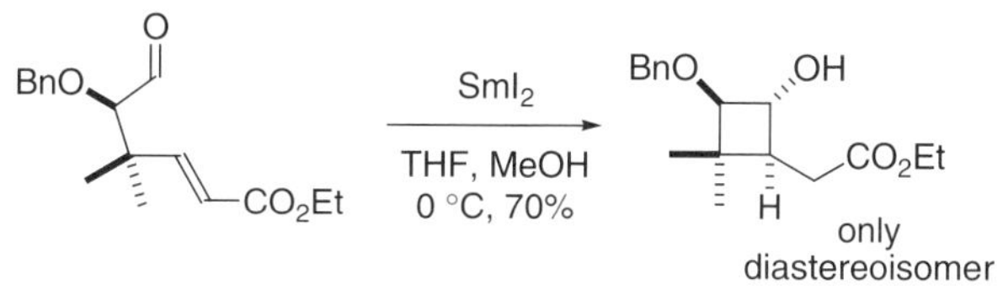

BnO O CO₂Et — SmI_2 / THF, MeOH, 0 °C, 70% → BnO OH CO₂Et H, only diastereoisomer

Scheme 5.31

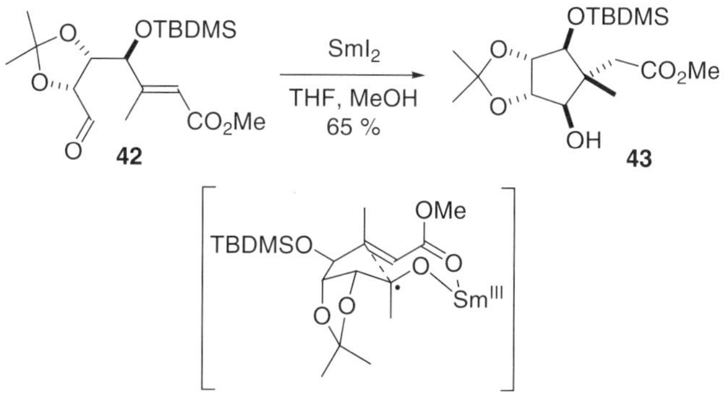

42 (OTBDMS, CO₂Me) — SmI_2 / THF, MeOH, 65 % → **43** (OTBDMS, CO₂Me, OH)

[TBDMSO, OMe, O, O, Sm^{III}, O O]

Scheme 5.32

Intramolecular SmI_2-mediated carbonyl–alkene couplings can also be used to construct challenging cage-like structures. For example, Nicolaou reported an impressive SmI_2-mediated cyclisation of aldehyde **44** as part of their total synthesis of the antibiotic platensimycin.[63] Alcohol **45** was formed after only 1 min using SmI_2 at −78 °C in a THF–HMPA mixture with hexafluoro-2-propanol (HFIP) (Scheme 5.33). An alternative mechanistic pathway involving initial reduction of the bisenone to the carbanion followed by nucleophilic addition to the aldehyde could also be operating in this case (see Scheme 5.19).

Representative procedure. To a stirred solution of SmI_2 in THF (2.2 equiv) at −78 °C was added HMPA (10% by volume of the THF used in reaction) and the colour of the solution changed from blue–green to purple–black. A solution of aldehyde **44** (1 equiv) and HFIP (1.5 equiv) in THF was added dropwise and the reaction mixture was stirred for 1 min. The reaction was quenched by opening to air and the addition of aqueous, saturated $Na_2S_2O_3$, followed by extraction with Et_2O. The organic extracts were combined, dried (Na_2SO_4) and the solvent removed *in vacuo*. Crude **45** was purified by column chromatography (EtOAc–hexane eluent).

A number of reports have appeared on the use of carbonyl–alkene additions to construct oxacyclic ring systems efficiently with high diastereoselectivity. Nakata showed that even the formation of seven-membered cyclic ethers using the transformation can be remarkably efficient and utilised the strategy to form the E ring of the polycyclic ether marine natural product yessotoxin (Scheme 5.34).[64]

An alternative means of promoting carbonyl–alkene cyclisations is to use activated alkenes in which the activating group can be removed spontaneously after cyclisation by elimination from an organosamarium intermediate. In 1997, Molander exploited this strategy in an elegant approach to enantiomerically enriched cyclopentanols.[65] In these reactions, SmI_2 promotes the asymmetric radical cyclisation of tartrate-derived keto allylic acetals such as **46** (Scheme 5.35). The acetal stereocontrol element gives remarkably good control at both reacting centres. Chelation to Sm(III) plays a central role in attaining the high levels of remote asymmetric induction and an ordered tricyclic transition structure involving a three-point chelation between the lanthanide metal ion of the ketyl radical anion and the acetal auxiliary has been proposed

Scheme 5.33

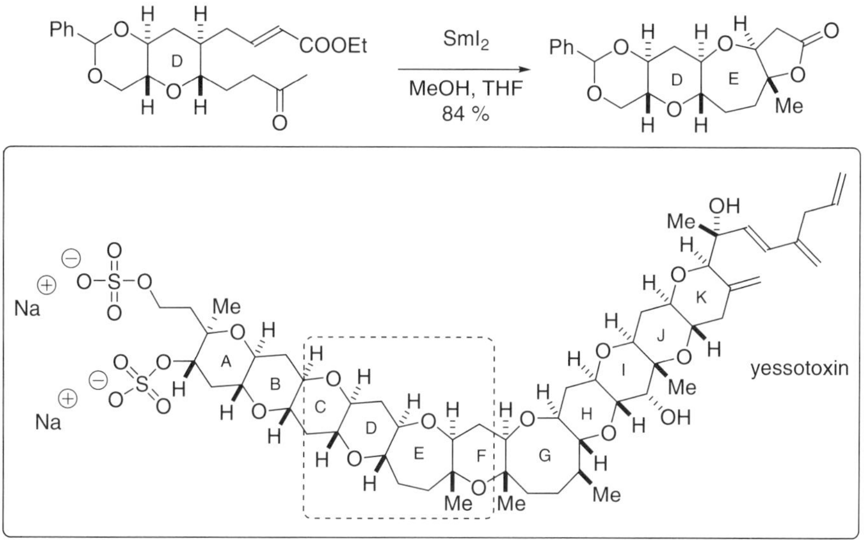

Scheme 5.34

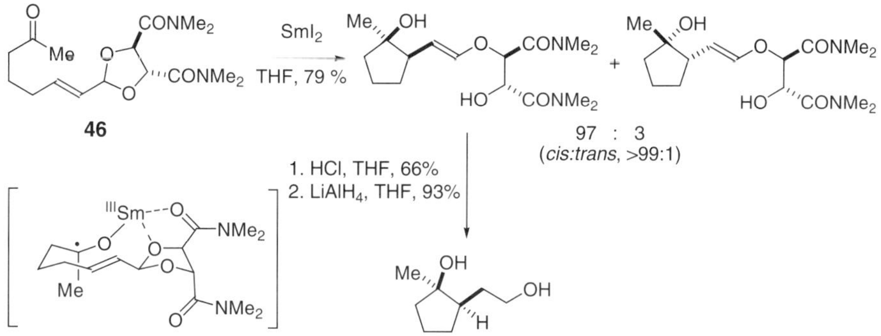

Scheme 5.35

(Scheme 5.35).[65] It is interesting that this is a rare example of a *syn*-selective carbonyl–alkene cyclisation.

Allylic sulfides can also be used as radical acceptors in carbonyl–alkene couplings, as demonstrated by Matsuda in his synthetic approach to grayanotoxin III.[23] After cyclisation, the spontaneous expulsion of sulfide installs a new double bond (see Chapter 7, Section 7.4).

Finally, Molander has shown that the efficiency of 8-*endo* carbonyl–alkene cyclisations can be significantly enhanced by the presence of a leaving group in the allylic position of the alkene.[57] For example, treatment of allylic acetate **47** with SmI_2 provides an excellent yield of the cyclooctenol **48** (Scheme 5.36). After the cyclisation step, the resultant secondary alkyl radical is reduced to an

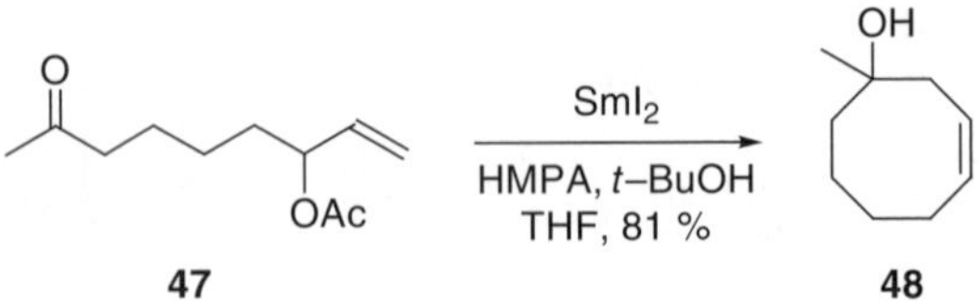

Scheme 5.36

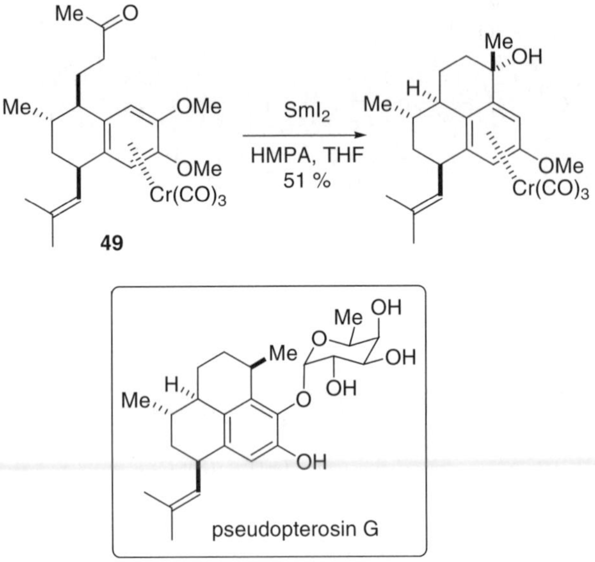

Scheme 5.37

organosamarium intermediate that undergoes β-elimination of the acetoxy group and generation of the endocyclic alkene.[57] The cyclisation of **47** provides a further illustration of the remarkable ability of SmI_2-mediated carbonyl–alkene couplings to form challenging cyclic systems when other radical reactions often prove inadequate.

5.2.4 Carbonyl–Arene Couplings

In addition to alkenes, arenes can sometimes be used as radical acceptors in SmI_2-mediated carbonyl–alkene couplings. For example, Schmalz reported extensive studies on ketyl additions to arenechromium tricarbonyl complexes:[66,67] tetralin–$Cr(CO)_3$ complex **49** underwent reductive carbonyl addition to the aromatic ring upon treatment with SmI_2 to furnish the skeleton of the naturally occurring aryl glycoside pseudopterosin G (Scheme 5.37).[66,67] Here, the bulky metal tricarbonyl group not only serves to control the stereochemistry of the cyclisation, but also activates the arene ring, which otherwise would be deactivated by the two electron-donating methoxy substituents. After radical addition to the aromatic system, reduction by a second SmI_2, protonation and subsequent rearomatisation takes place by regioselective elimination of one of the methoxyl groups.

Representative procedure. To a stirred solution of SmI_2 in THF (2.4 equiv) at room temperature was added HMPA (18 equiv) and the colour of the solution changed from blue–green to purple–black. A solution of ketone (1 equiv) and *tert*-butanol (2 equiv) in THF (0.06 M) was added dropwise and the reaction mixture was stirred for 16 h at room temperature. The reaction was quenched by opening to air and the addition of aqueous, saturated $NaHCO_3$, followed by extraction with Et_2O. The organic extracts were combined, dried (Na_2SO_4) and the solvent removed *in vacuo*. The crude product was purified by column chromatography (EtOAc–hexane eluent).

Arene activation by metal complexation is not always necessary for the cyclisation of samarium ketyls on to aromatic systems. Studies carried out by Reissig demonstrated the utility of these reactions for the synthesis of a wide variety of polycyclic systems, although yields and stereoselectivities depend highly on the substrate structure.[68] One well-behaved system is represented by the cyclic γ-naphthyl ketones **50–54**, which cyclise with excellent diastereocontrol, thereby incorporating four- to eight-membered rings into the tetracyclic structures **55–59** (Scheme 5.38).[68] Tetracyclic product **56** ($n = 1$) is of particular interest because of its steroid-like structure.

Reissig also described SmI_2-induced cyclisations of indole derivatives in a stereoselective synthesis of highly functionalised benzannulated pyrrolizidines and indolizidines.[69] After the cyclisation of samarium ketyl radical **61**, a second electron transfer gives a samarium enolate **62** that can be quenched with carbon electrophiles such as allyl bromide to give products such as **63**. These sequential reactions selectively generate three contiguous stereogenic centres including a challenging quaternary centre at the 3-position of the indole moiety – a structural motif found in many indole alkaloids (Scheme 5.39).[69]

SmI2
HMPA, t–BuOH
THF, rt

50 n = 0
51 n = 1
52 n = 2
53 n = 3
54 n = 4

55 n = 0, 41 %
56 n = 1, 80 %
57 n = 2, 96 %
58 n = 3, 98 %
59 n = 4, 52 %

Scheme 5.38

Scheme 5.39

Representative procedure. To a stirred solution of SmI_2 in THF (2.5 equiv) at room temperature was added HMPA (10 equiv) and the colour of the solution changed from blue–green to purple–black. A solution of *N*-acylated indole derivative **60** (1 equiv) in THF was added dropwise and the reaction mixture was stirred for 16 h. Allyl bromide (no equivalents given) was added and the reaction mixture was stirred (no reaction time specified). The reaction was quenched by opening to air and the addition of aqueous, saturated $NaHCO_3$, followed by extraction with Et_2O. The organic extracts were combined, dried (Na_2SO_4) and the solvent removed *in vacuo*. Crude **63** was purified by column chromatography (EtOAc–hexane eluent).

5.2.5 Reductive Couplings of Imines and Their Equivalents

As for SmI_2-mediated pinacol couplings (see Section 5.1.4), imine equivalents have recently been used in place of the carbonyl component in these reductive couplings. In particular, the groups of Skrydstrup and Py have shown that nitrones are excellent coupling partners in these reactions. In work on the asymmetric synthesis of γ-amino acids, Skrydstrup studied the reductive coupling of nitrones and α,β-unsaturated esters bearing a chiral auxiliary, leading to the formation of *N*-hydroxyamino acids in good yields.[70] The best diastereoselectivities were achieved when nitrone **64** was coupled with the acrylate of (1*R*,2*S*)-*N*-methylephedrine **65** (Scheme 5.40).[70]

The use of a nitrone equivalent of a carbonyl–alkene coupling also provides an alternative approach for asymmetric control through the attachment of a chiral auxiliary to the nitrogen of the nitrone. This strategy is clearly not possible for analogous reactions involving aldehydes and ketones. Skrydstrup demonstrated the utility of the SmI_2-promoted reductive addition of alkyl nitrones, bearing a carbohydrate-based auxiliary, to *n*-butyl acrylate, providing addition products in good yield and in high diastereoisomeric excess (Scheme 5.41).[71] A simple acid-mediated hydrolysis step removes the auxiliary and liberates the

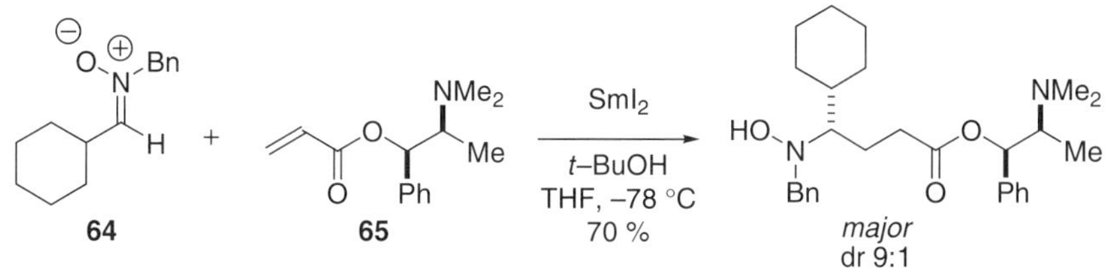

Scheme 5.40

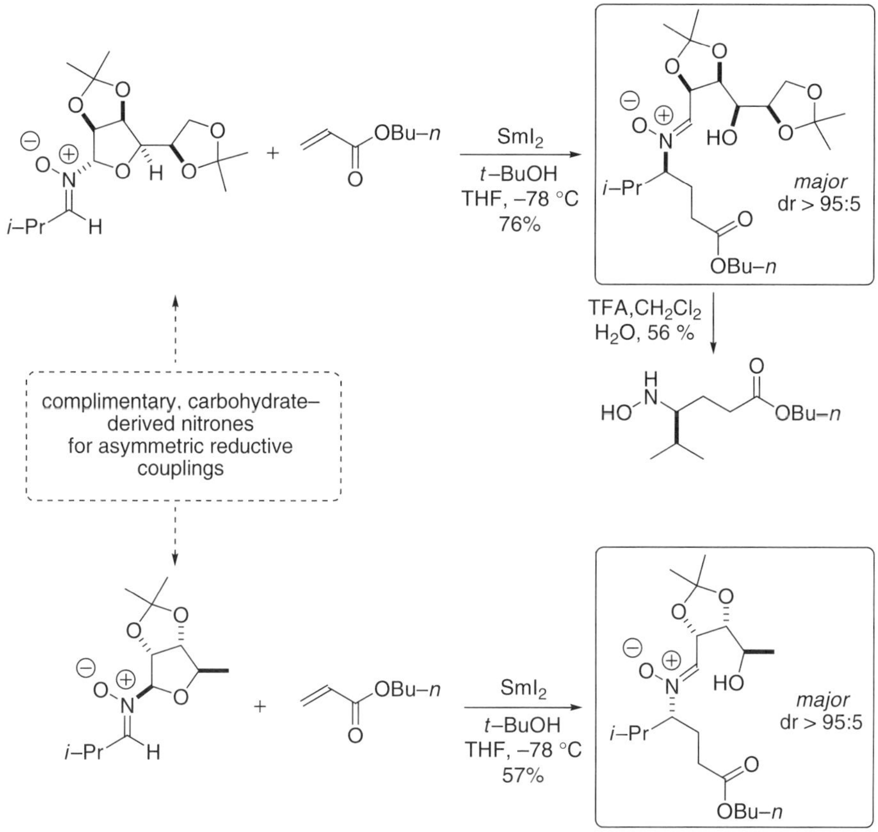

Scheme 5.41

enantioenriched *N*-hydroxy-γ-amino acids. Both enantiomers of the amino acid derivatives can be obtained by the judicious choice of the carbohydrate auxiliary. A chelated transition structure was postulated to explain the stereochemical outcome of these reductive addition reactions.[71]

Representative procedure. To a stirred solution of nitrone (1 equiv) and acrylate (1 equiv) in THF (0.05 M) at −78 °C was added *tert*-butanol (2 equiv) followed by dropwise addition of a solution of SmI_2 in THF (2.4 equiv) and the reaction mixture was stirred for 20 h. The reaction was quenched by opening to air and the addition of aqueous, saturated NH_4Cl with subsequent dilution with water, followed by extraction with CH_2Cl_2. The organic extracts were combined, washed with brine, dried (Na_2SO_4) and the solvent removed *in vacuo*. The crude product was purified by column chromatography (EtOAc–pentane eluent).

An alternative chiral auxiliary, the 1-(triisopropylphenyl)ethyl group, was successfully used by Py for similar asymmetric reductive couplings (Scheme 5.42).[72,73] It is noteworthy that these SmI_2-mediated coupling reactions also gave high diastereoselectivities when α- and β-substituted enoates were used as the coupling partners.

Py used a nitrone–alkene coupling in a short, asymmetric synthesis of (+)-hyacinthacine A_2 (Scheme 5.43).[74] The key carbon–carbon bond-forming step in their approach involved the highly diastereoselective reductive coupling of the L-xylose-derived cyclic nitrone **66** to ethyl acrylate.[74]

Finally, in 2006, Ellman reported a single example of a sulfinimine–alkene coupling mediated by SmI_2 in their work on the total synthesis of the anticancer agent, tubulysin D.[75] It was found that the coupling of enantiomerically pure sulfinimine **67** with methyl methacrylate was effectively promoted by the

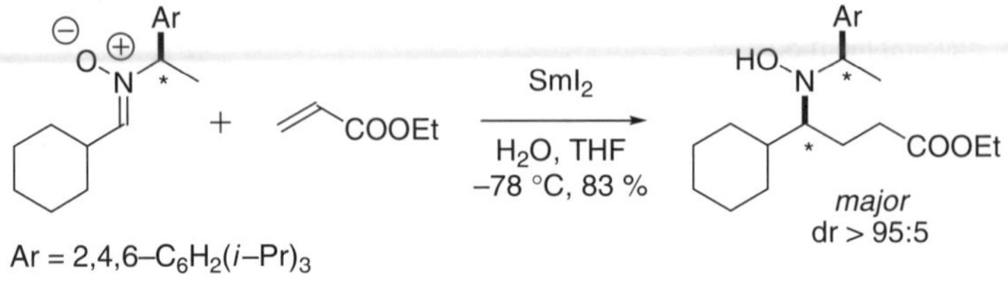

Ar
+
COOEt
SmI_2
H_2O, THF
–78 °C, 83 %
HO
N
Ar
COOEt
major
dr > 95:5
Ar = 2,4,6–$C_6H_2(i–Pr)_3$

Scheme 5.42

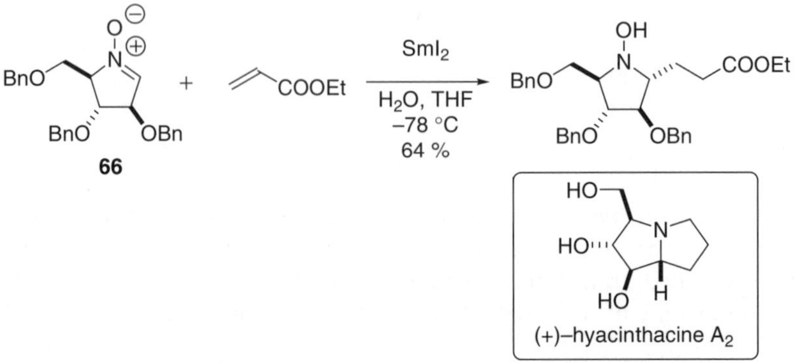

BnO
BnO
OBn
66
+
COOEt
SmI_2
H_2O, THF
–78 °C
64 %
OH
N
COOEt
BnO
BnO
OBn
HO
HO
N
HO
H
(+)–hyacinthacine A_2

Scheme 5.43

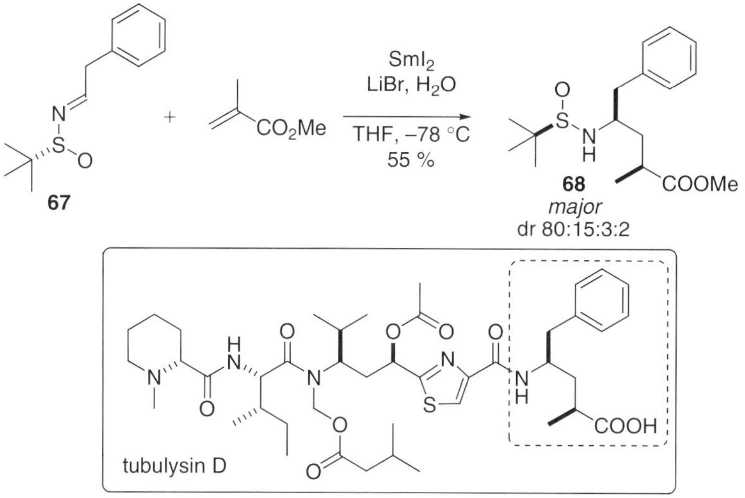

Scheme 5.44

presence of water and LiBr to give the desired adduct **68** with good diastereocontrol (Scheme 5.44).[75]

5.2.6 Conclusions

The reductive coupling of carbonyls and alkenes has become one of the most popular carbon–carbon bond-forming reactions mediated by SmI_2. No other reagent can perform this type of reaction with such efficiency and versatility. In many instances, the reactions provide products with high levels of diastereocontrol that can be ascribed to the hard Lewis acidic Sm(III) counter ion of the metal ketyl intermediate providing well-defined transition structures through chelation to the reacting partners. The many applications of this transformation in the total synthesis of complex targets are a testament to its utility.

5.3 Radical–Alkene/Alkyne Additions

5.3.1 Introduction

The addition of sp^3 or sp^2 carbon-centred radicals to unsaturated carbon–carbon bonds is yet another class of synthetically useful reactions promoted by the low-valent lanthanide reagent SmI_2. Halides and sulfones are the most common functional groups used as precursors to radicals, although other groups have also been successfully employed. Although the intermolecular variant of this reaction has found only limited application, intramolecular variants can be highly successful. In most of these cyclisation events, the

$$R{-}X \xrightarrow{SmI_2} R^{\bullet} \begin{cases} \xrightarrow[\text{reduction}]{k_{red}} R{-}SmI_2 \\ \xrightarrow[\text{radical cyclisation}]{k_{cyc}} R'^{\bullet} \longrightarrow R'{-}SmI_2 \end{cases}$$

X = halide (or sulfone)

Scheme 5.45

coupling partners are unactivated alkenes and alkynes, but the methodology can also be extended to include other radical acceptors such as α,β-unsaturated amides, esters, nitriles, lactams and lactones.

The use of SmI_2 has clear advantages over the use of tin hydride reagents to initiate radical cyclisations. First, lower reaction temperatures can be used in conjunction with SmI_2, as a thermally labile initiator is not required to generate the radical as it is when tributyltin hydride is used. Second, the lanthanide salts generated after completion of the reaction are considerably less toxic than the tin by-products produced from tributyltin hydride and they can be easily separated from the desired cyclic products. On the other hand, there are also intrinsic limitations to be considered in SmI_2-mediated radical cyclisations. As discussed in detail in Chapter 3, Section 3.2, the conditions for generating an intermediate alkyl radical are also the same as those for the preparation of the organosamarium species through an additional single-electron reduction step (Scheme 5.45). Hence, for successful ring closure, the unimolecular rate constant for cyclisation (k_{cyc}) must be sufficiently higher than the bimolecular rate constant for the second reduction step multiplied by the SmI_2 concentration in the reaction mixture ($k_{red}[SmI_2]$).[76] Nevertheless, considering that the cyclisations are indeed unimolecular events, performing the reactions at low substrate concentrations can impede premature reduction of the initial radical formed.

To date, the 5-*exo-trig* and 5-*exo-dig* radical cyclisations represent the most successful classes of SmI_2-mediated radical-alkene/alkyne cyclisations and have been applied to a diverse array of substrates. The application of this approach to the generation of larger ring sizes is more challenging and success is highly dependent on the structure of the radical precursor.

5.3.2 Radical Additions to Alkenes

Alkyl radicals generated from the reduction of halides or sulfones with SmI_2 have been successfully exploited in intramolecular additions to alkenes that result in the generation of a variety of functionalised small carbocyclic and heterocyclic ring systems. Substrates containing an oxygen atom within the

n–BuO O I

69

SmI_2

HMPA, THF
rt, 2 h

n–BuO O

H_2CrO_4
70% overall

O O

major
dr 98:2

Scheme 5.46

tether linking the radical and the alkene undergo particularly successful 5-*exo-trig* cyclisations with rates that are approximately an order of magnitude faster than those observed for the corresponding all-carbon analogue. For example, Fukuzawa showed that unsaturated alkyl halides, such as **69**, undergo efficient cyclisation in an efficient two-step synthesis of γ-lactones (Scheme 5.46).[77] It is noteworthy that the radical cyclisation is faster than the reduction of the primary radical intermediate, which would have led to β-elimination of one of the two alkoxy groups.

Similar examples involving glycosyl sulfones have been exploited in the stereoselective synthesis of *C*-glycosides. Sinaÿ found that the glycosyl phenyl sulfone **70** underwent cyclisation upon treatment with SmI_2–HMPA to give the bicyclic product **71** as a mixture of diastereoisomers, although complete stereocontrol was achieved at the anomeric centre (Scheme 5.47).[78] Skrydstrup and Beau subsequently used a silicon tether for the preparation of simple alkyl *C*-glycosides from substrates such as the mannosyl 2-pyridyl sulfone **72** (Scheme 5.47).[79] Reduction of the corresponding phenyl sulfone with SmI_2 could be carried out with HMPA as an additive, although the yields were unsatisfactory. Lowering the LUMO of the aryl sulfone by switching to the 2-pyridyl sulfone system, however, allowed the cyclisation event to proceed efficiently in the absence of a coadditive, furnishing the *C*-glycoside **73** in good yield after desilylation (Scheme 5.47).[80]

Electron-deficient alkenes, such as α,β-unsaturated esters and amides (and alkynes), can also be exploited in SmI_2-mediated radical cyclisations. Molander reported the use of such ring closures for the generation of a variety of monocyclic and bicyclic carbocycles in good yield and with excellent diastereoselectivity (Scheme 5.48). The yields and stereoselectivities often exceed those obtained in similar radical cyclisations using alternative reagents.[81] The addition of HMPA was sometimes detrimental to these cyclisations, whereas alternative additives such as NiI_2 proved beneficial by shortening reaction times. These additions typically require the presence of a proton source (MeOH or *t*-BuOH) to quench the Sm(III) enolate that results from the addition, thus preventing side reactions (Scheme 5.48).[81]

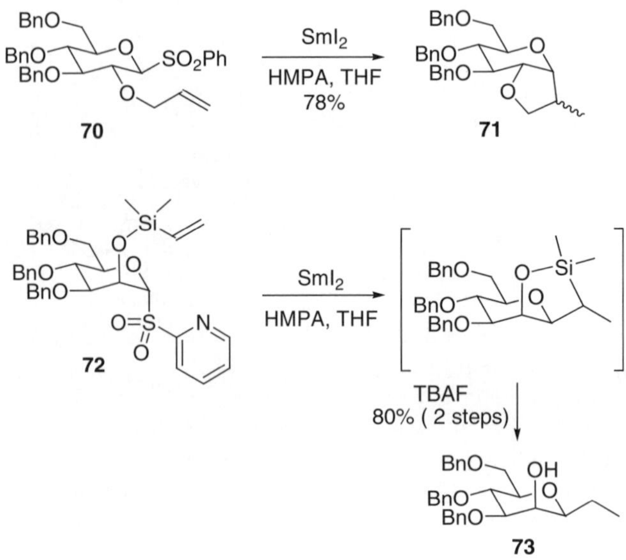

Scheme 5.47

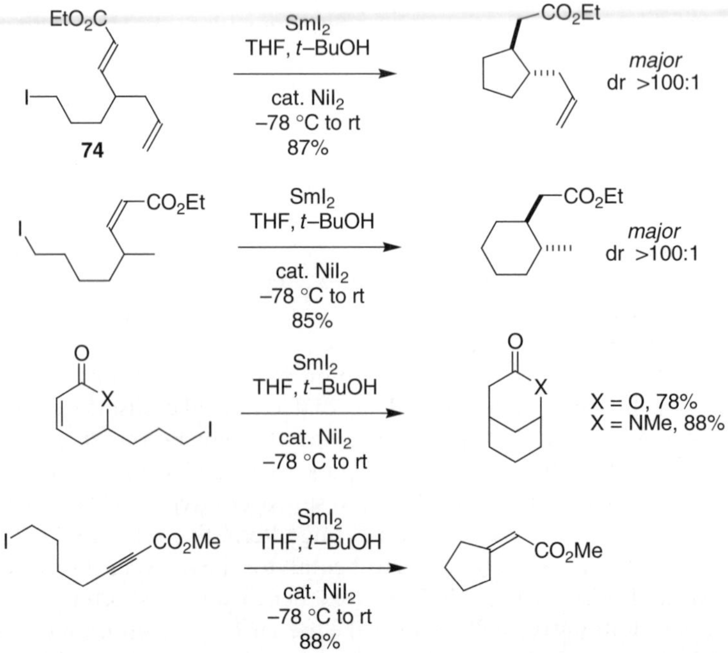

Scheme 5.48

Representative procedure. To Sm metal (0.25 mmol) in dry THF (14 ml) was added CH_2I_2 (1.5 mmol) and the mixture was stirred at room temperature for 1.5 h. NiI_2 (cat.) was added and the solution then cooled to −78 °C before addition of the substrate (0.5 mmol) and *t*-BuOH (1 mmol). The reaction mixture was stirred for 30 min, then allowed to warm to room temperature, where TLC analysis showed complete consumption of the starting material. The reaction was quenched with aqueous saturated $NaHCO_3$, filtered through Celite and the aqueous layer extracted. The organic layers were combined and the solvent was removed *in vacuo* to yield the crude product that was purified by either Kugelrohr distillation or column chromatography (EtOAc–hexane eluent).

The high *trans* diastereoselectivity often observed in cyclisations involving substrates bearing substituents at the C4 position, such as **74**, can be rationalised by comparing the two possible chair-like transition structures (Scheme 5.49). Positioning the R group substituent in a pseudo-equatorial orientation relieves the $A^{1,3}$ strain observed for the alternative conformer with the same substituent in a pseudo-axial position.[81]

Bennett has shown that sugar-derived substrates can also be used for the preparation of highly functionalised cyclopentanes.[82,83] For example, treatment of iodide **75** with SmI_2 initiated a highly diastereoselective construction of cyclopentane **76** (Scheme 5.50). These reactions require the presence of HMPA and a proton source to avoid competing 1,4-reduction of the α,β-unsaturated ester and cleavage of the allylic C–O bond.[82,83]

Representative procedure. The substrate (0.35 mmol) was dissolved in THF (9.2 ml) and MeOH (0.14 ml) and cooled to −78 °C under an argon atmosphere. In a separate flask, HMPA (1.2 ml) and a 0.1 M solution of SmI_2 (1.4 mmol) was stirred for 10 min before being added to the substrate using a cannula. The reaction mixture was stirred at −78 °C and then 0 °C before being quenched with aqueous

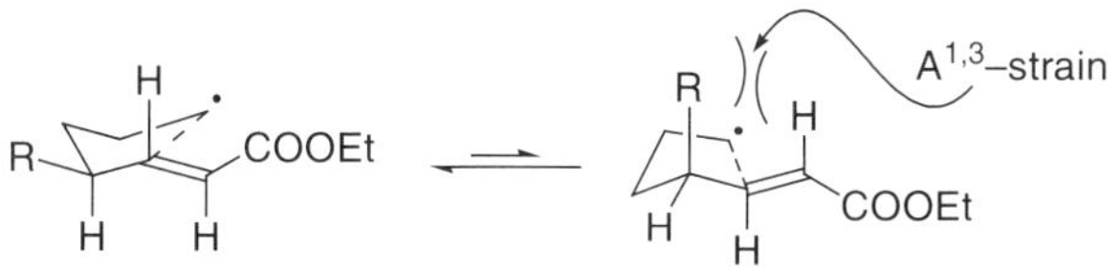

Scheme 5.49

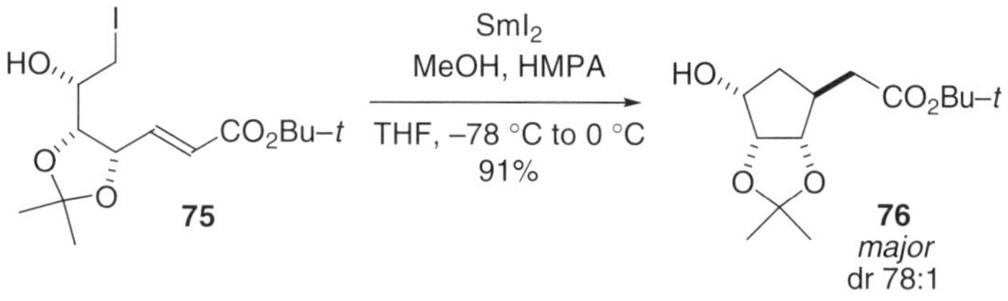

Scheme 5.50

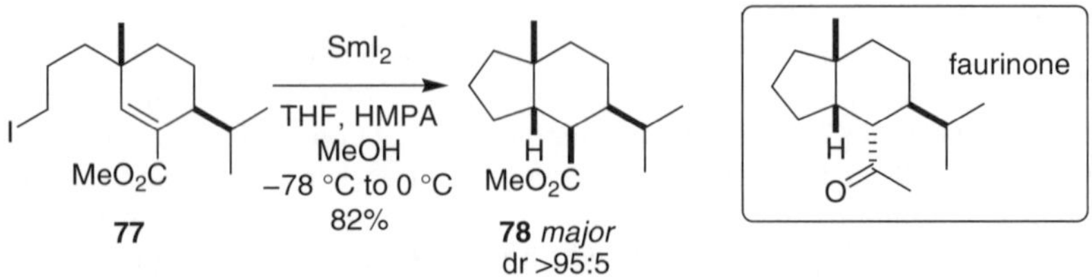

Scheme 5.51

acid. The aqueous layer was extracted with EtOAc and the combined organic layers were washed with H_2O, aqueous saturated $Na_2S_2O_3$ and brine. The combined organic layers were dried using $MgSO_4$ and the solvent was removed *in vacuo*. The crude product was purified by column chromatography (EtOAc–hexane eluent).

Procter employed a highly diastereoselective SmI_2-mediated intramolecular 1,4-addition during studies on the synthesis of faurinone: treatment of iodide **77** with SmI_2 in THF–HMPA gave *cis*-hydrindane **78** in good yield (Scheme 5.51).[84]

Representative procedure. The substrate (0.12 mmol) and MeOH (1.24 mmol) in THF (2 ml) were added dropwise to a 0.1 M solution of SmI_2 (0.5 mmol) and HMPA (2.48 mmol) at −78 °C. The solution was warmed to 0 °C and stirred for 4 h, then quenched with aqueous saturated NH_4Cl. The aqueous layer was extracted with Et_2O and the organic layers were combined, dried ($MgSO_4$) and the solvent removed *in vacuo*. The crude product was purified by column chromatography (Et_2O–hexane eluent).

N-[(*N′*,*N′*-Dialkylamino)alkenyl]benzotriazoles have been shown to be useful precursors to α-amino radicals, which can subsequently undergo both 5-*exo*- and 6-*exo* ring closure on to α,β-unsaturated esters or nitriles.[85] For example, exposure of benzotriazole **79** to SmI_2 afforded *cis*-cyclopentane **80** in 70% yield (Scheme 5.52). The mechanism of the cyclisation is believed to involve dissociation of the benzotriazole, thereby generating an iminium cation, which undergoes rapid reduction to the α-amino radical in the presence of SmI_2 (Scheme 5.52).[85]

Aryl halides have also been widely exploited in 5-*exo-trig* and 6-*exo-trig* radical cyclisations. For example, the ABC ring system of the natural product aflatoxin B_1 was constructed by Holzapfel using a SmI_2-promoted cyclisation of the aryl iodide **81**, followed by reduction of the product radical and subsequent elimination of the acetate group (Scheme 5.53).[86] Lu and Cai reported a synthesis of the natural products cryptotanshinone and tanshinone IIA, in which the dihydrofuran ring in the intermediate **83** was installed using an efficient SmI_2-mediated cyclisation of the aryl bromide **82** (Scheme 5.53).[87]

5.3.3 Radical Additions to Alkynes

Carbon-centred radicals generated by reductions with SmI_2 undergo intramolecular additions to alkynes, typically via 5-*exo-dig* pathways, providing

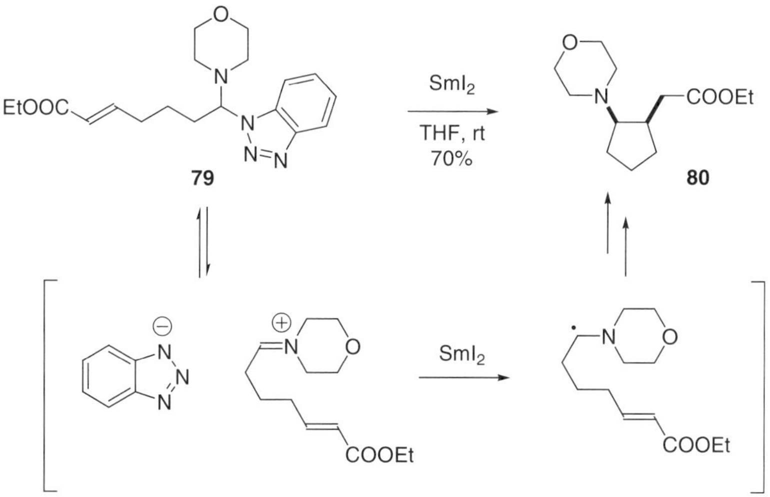

Scheme 5.52

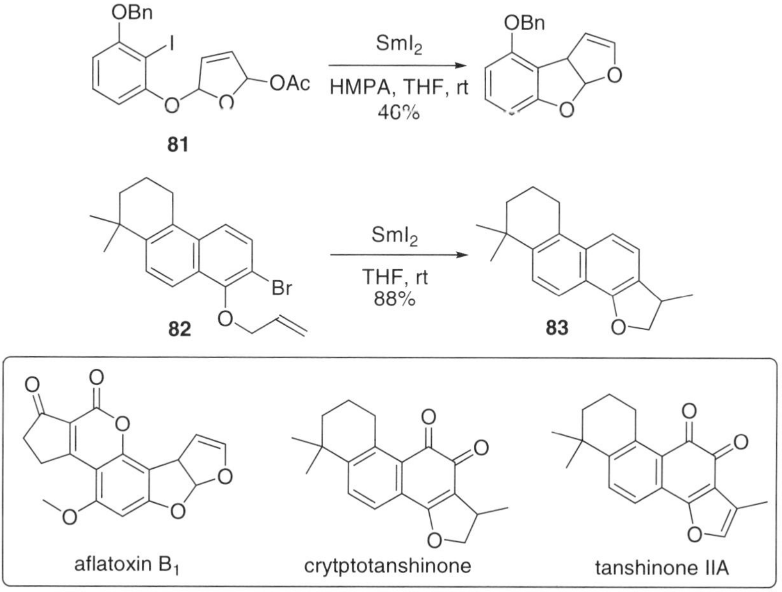

Scheme 5.53

cyclic systems bearing exomethylene substituents. In contrast to analogous 5-*exo* cyclisations involving alkenes, the alkenyl radicals formed after ring closure have two possible fates depending on the nature of the alkenyl radical: Unstabilised alkenyl radicals abstract a hydrogen atom from the solvent faster than their reduction to an sp^2 carbanion, whereas radicals stabilised by neighbouring groups, such as arenes or silyl groups, are reduced to the corresponding anion and protonation by an additive becomes the major pathway.[79,88,89]

This type of SmI_2-mediated cyclisation has been used in the construction of heterobicyclic systems found in natural products. For example, Ohta constructed the cyclopentane ring of oxerine by a SmI_2-promoted 5-*exo-dig* cyclisation of the bromide **84**, generating the fused ring system **85** in high yield (Scheme 5.54).[90] No products from the elimination of the benzyloxy substituent were observed, attesting to the radical nature of this ring-closing step. The same group exploited a similar cyclisation strategy for the synthesis of pyrrolam A.[91] Treatment of bromide **86** with SmI_2 in the presence of HMPA triggered a 5-*exo-dig* cyclisation to give the bicyclic lactam **87** in 90% yield (Scheme 5.54).

Aryl and alkenyl radicals also undergo cyclisation on to alkynes, although such ring closures are less effective than examples that involve an initial sp^3 carbon-centred radical (see Scheme 5.54). Inanaga reported the application of aryl halide–alkyne cyclisations in the construction of indole and benzofuran frameworks.[92] For example, treatment of propargylamine **88** with SmI_2 in THF–HMPA gave 3-substituted indole **89** in moderate yield after cyclisation and double bond isomerisation. Alternatively, the cyclisation of the aryl iodide **90** with SmI_2 gave benzofuran **91**, possessing an exocyclic double bond (Scheme 5.55).[93]

Representative procedure. To the substrate (0.15 mmol) in THF at room temperature was added a 0.1 M solution of SmI_2 (0.45 mmol) and HMPA (0.4 ml). The reaction was quenched, worked up and the crude mixture purified by column chromatography.

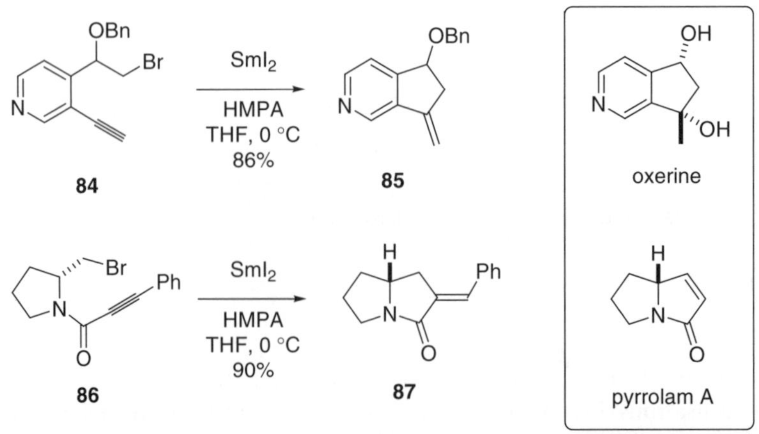

Scheme 5.54

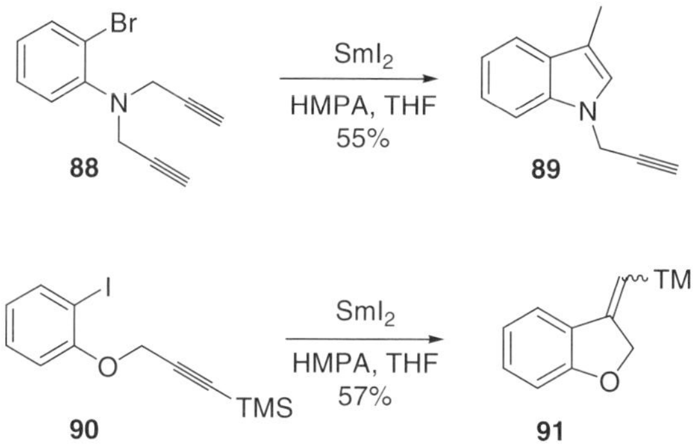

Scheme 5.55

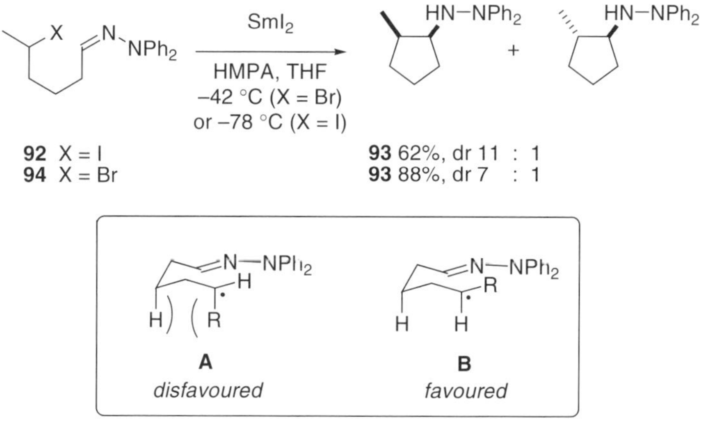

Scheme 5.56

5.3.4 Radical Additions to Other Unsaturated Groups

As discussed in Section 5.1, imines and imine equivalents can also function as effective radical acceptors. Hydrazones have been shown to be highly effective acceptors for alkyl radicals, with rate constants for 5-*exo*-cyclisations in the range of two orders of magnitude higher than those of the corresponding 5-hexenyl radical cyclisations.[26,94] For example, treatment of the secondary iodide **92** with SmI_2–HMPA gave cyclopentylhydrazine **93** with good diastereoselectivity (Scheme 5.56). Although higher cyclisation yields were achieved using the corresponding bromide **94**, the *cis*-diastereoselectivity was lower as a result of the higher reaction temperatures required for the initial reduction step.[26,94]

The *cis*-selectivity of these reactions can be rationalised by invoking a pseudo-chair transition structure. Transition structure **A** leading to the *trans* product suffers from a 1,3-diaxial interaction between the alkyl substituent and an axial hydrogen atom on the ring. This unfavourable interaction is

SmI_2

HMPA, THF

95 R = Me
96 R = Bn

72%, dr 5 : 1
78%, dr 9 : 1

Scheme 5.57

absent in a similar transition structure **B** leading to the *cis* product (Scheme 5.56).[26,94]

Oximes are equally viable radical acceptors in SmI_2-mediated radical cyclisations. For example, carbohydrate-derived oxime substrates **95** and **96** undergo cyclisation upon treatment with SmI_2 to give *N*-alkoxyaminocyclopentanes in good yield with satisfactory diastereoselectivity (Scheme 5.57).[95,96] As in the analogous cyclisations of hydrazones, the efficiency of these reactions arises from the stabilisation of the aminyl radical, resulting from the cyclisation, by the lone pair of the adjacent heteroatom.

Representative procedure. To a solution of the oxime substrate (0.3 mmol) in THF (10 ml) and HMPA (1 ml) at 0 °C was added a 0.1 M solution of SmI_2 (0.9 mmol) over 5 min. The reaction mixture was allowed to warm to room temperature for 30 min, then quenched with aqueous saturated NH_4Cl solution. The reaction mixture was diluted with H_2O and extracted with EtOAc and the combined organic layers were washed with H_2O, aqueous saturated $Na_2S_2O_3$ and brine. The combined organic layers were dried using $MgSO_4$ and the solvent was removed *in vacuo*. The crude product was purified by column chromatography (EtOAc–hexane eluent).

5.3.5 Conclusions

SmI_2-mediated radical cyclisations involving alkyl, alkenyl and aryl radical intermediates can be used to construct efficiently five-membered and, in certain cases, six-membered ring systems. This approach provides a useful alternative to trialkyltin hydride-mediated methods as toxic reagents and problematic tin by-products are avoided. In addition, the use of SmI_2 to induce radical cyclisations has led to the development of a number of powerful, radical/anionic sequential processes for the construction of complex systems. Sequential reactions involving radical–alkene/alkyne cyclisations are discussed in Chapter 6.

5.4 SmI_2-mediated Barbier and Grignard Reactions

5.4.1 Introduction

In Chapter 4, we saw that SmI_2 mediates the dehalogenation of a range of alkyl halides. These reactions proceed by reduction to alkyl radicals that are then reduced further by SmI_2 to give organosamarium intermediates. The organosamariums are then quenched to give the corresponding alkanes. The intermediates generated in these reductions can be exploited in reactions with carbonyl groups. These reactions are referred to as SmI_2-mediated Barbier or Grignard reactions. The 'Barbier' and 'Grignard' labels are used to differentiate between *inter*molecular procedures where a mixture of the alkyl halide and carbonyl compound is treated with SmI_2 (Barbier) and where the alkyl halide is treated with SmI_2 prior to the addition of the carbonyl compound (Grignard) (Scheme 5.58). The SmI_2-mediated Barbier reaction has been the subject of an extensive review by Krief and Laval.[97]

As organosamariums are formed from radical intermediates, it is important to note that competing radical processes may be problematic if they proceed at a rate that is greater than the rate of reduction of the intermediate radical to the organosamarium (see Chapter 3, Section 3.2). In general, Barbier conditions should be used when the organosamarium intermediate is prone to dimerisation (*e.g.* allylic and benzylic organosamariums), decomposition (*e.g.*

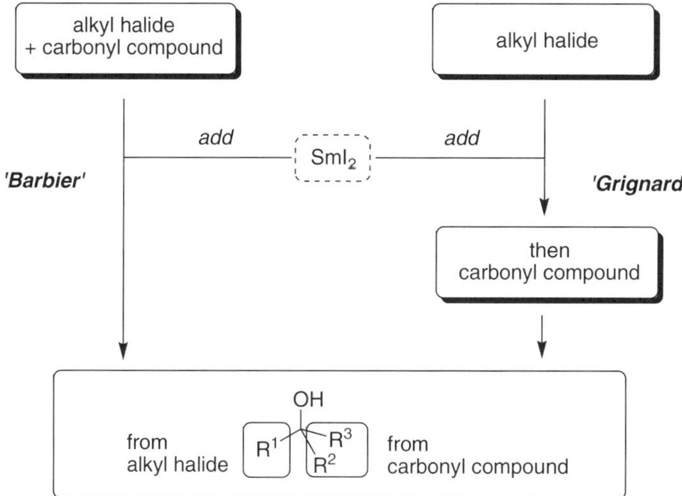

'**Barbier** conditions' should be considered for:

- allylic and benzylic halides
- alkyl halides that lead to unstable organosamariums
- alkyl halides that lead to organosamariums that can react with themselves

'**Grignard** conditions' should be considered for:

- carbonyl compounds that can be readily reduced
- alkyl halides that reduce slowly
- reactions in which transmetallation of the organosamarium is required

Scheme 5.58

by elimination) or self-condensation. Grignard conditions suit reactions in which the carbonyl compound may be reduced faster than the alkyl halide, thus leading to competing reactions (*e.g.* aryl ketones and aldehydes) (Scheme 5.58).[97] The expected order of reactivity is observed in SmI_2-mediated Barbier and Grignard reactions (RI > RBr > > RCl): alkyl iodides and bromides undergo facile reaction whereas alkyl chlorides typically do not react. (Allylic chlorides do react although they do so slowly.)[98] Tosylates have been converted to iodides *in situ* using NaI and have also been employed in Barbier couplings (Scheme 5.59).[98]

SmI_2-mediated Barbier reactions are typically carried out in THF with additives, such as HMPA,[10,99] NiI_2[100–102] and ferric salts such as $Fe(DBM)_3$ (DBM=dibenzoylmethido).[98,100,103] Light activation[101,102] has also been used to increase the rate of reactions. Flowers has shown that SmI_2-mediated Barbier reactions in the presence of HMPA proceed through outer-sphere electron-transfer processes.[104] Although the role of metal salt additives is not clear, there are often clear benefits from their use. For example, Molander has shown that the use of NiI_2 in conjunction with visible-light activation allows the challenging addition of alkyl chlorides to esters to be carried out.[101–102,105] NiI_2 has also been used to catalyse the SmI_2-mediated addition of alkyl iodides to nitriles.[106]

The SmI_2-mediated Barbier reaction is homogeneous and often highly chemoselective, which is an advantage over Barbier reactions mediated by other metals (*e.g.* Mg, Li, Zn). While primary, secondary, allylic and benzylic halides can be used in the transformation in THF, the use of THP as solvent is often beneficial.[107] Aryl, vinyl and alkynyl halides can also be used in SmI_2-mediated Barbier reactions; however, although these organohalides are reduced to the corresponding radicals by SmI_2, hydrogen atom abstraction from THF is typically faster than further reduction to the required organosamarium.[108] Therefore, for aryl, vinyl and alkynyl halides, benzene is used as the solvent for reactions with SmI_2–HMPA.[109] Organosamariums typically undergo 1,2-addition to enones rather than 1,4-addition unless their reactivity is

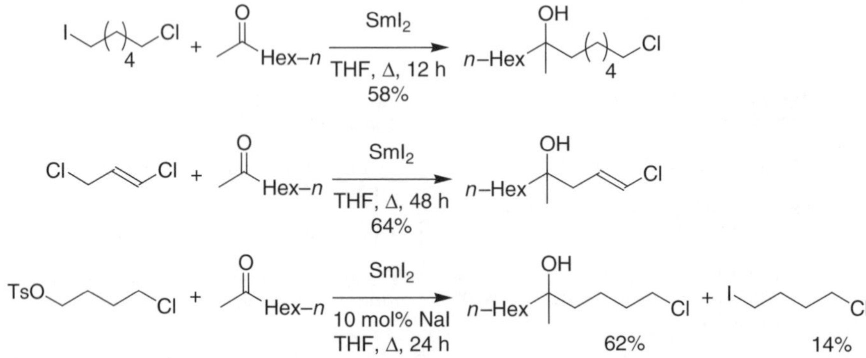

Scheme 5.59

Table 5.1 Conditions employed in SmI_2-mediated Barbier reactions for alkyl halides.

alkyl halide (or equivalent)	*additives*	*solvent*	*carbonyl electrophile*
alkyl iodide	none, HMPA Fe(III), Ni(II)	THF	aldehydes, ketones, esters, amides, imides, nitriles
alkyl bromide	none, HMPA	THF	aldehydes, ketones, esters, amides (imides)
alkyl chloride	HMPA, Fe(III) Ni(II) and h_ν	THF	aldehydes, ketones (esters)
alkyl tosylate	NaI	THF	aldehydes, ketones
allylic and benzylic iodide	none	THF, THP	aldehydes, ketones
allylic and benzylic bromide	none	THF, THP	aldehydes, ketones
allylic and benzylic chloride	none, HMPA	THF, THP	aldehydes, ketones (esters)
aryl halide, vinyl halide	HMPA	benzene, THF	ketones (and aldehydes)
alkynyl halide	HMPA	benzene, THF	ketones (and aldehydes)

modified by transmetallation.[110] It should also be noted that Barbier and Grignard reactions can be carried out using catalytic SmI_2 and mischmetal as the stoichiometric reductant (see Chapter 7, Section 7.2).[111] Table 5.1 summarises the additives, solvents and carbonyl electrophiles typically employed in SmI_2-mediated Barbier reactions for particular classes of alkyl halide.

5.4.2 The Mechanism of SmI_2-mediated Barbier and Grignard Reactions

SmI_2-mediated Barbier and Grignard reactions permit the coupling of alkyl halides and carbonyl compounds to give alcohols (or ketones if esters are employed as the electrophiles). When an intermolecular coupling is carried out under Grignard conditions, Curran has shown the mechanism to be straightforward: the alkyl halide is reduced by 2 equiv of SmI_2, *via* alkyl radicals, to give organosamarium species that react with carbonyl electrophiles in a subsequent stage of the reaction.[112] The organosamariums formed under Grignard conditions using SmI_2 are typically not long-lived and must be used quickly.

However, ambiguity surrounds the mechanism of the *inter*molecular SmI_2-mediated Barbier reaction, as two reducible functional groups, the alkyl halide and the carbonyl group, are present in the reaction mixture with SmI_2. In 1992, after a thorough review of the mechanistic evidence, Curran proposed that the SmI_2-mediated Barbier reaction also involves the addition of an organosamarium or carbanion to a carbonyl group and this mechanism has become widely accepted (Scheme 5.60).[108,112]

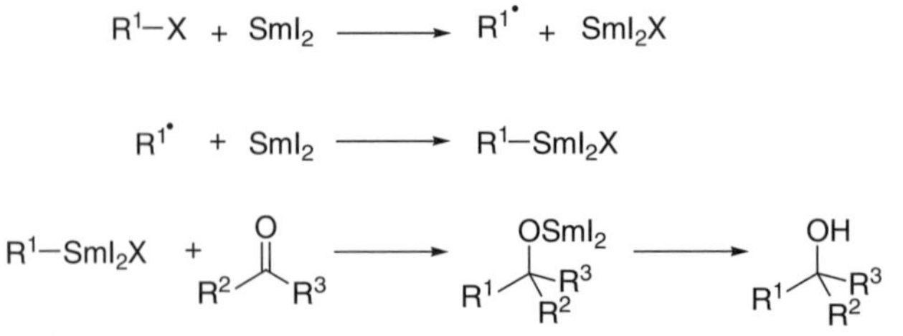

$R^1{-}X + SmI_2 \longrightarrow R^{1\bullet} + SmI_2X$

$R^{1\bullet} + SmI_2 \longrightarrow R^1{-}SmI_2X$

$R^1{-}SmI_2X + R^2C(O)R^3 \longrightarrow R^1R^2R^3C{-}OSmI_2 \longrightarrow R^1R^2R^3C{-}OH$

Scheme 5.60

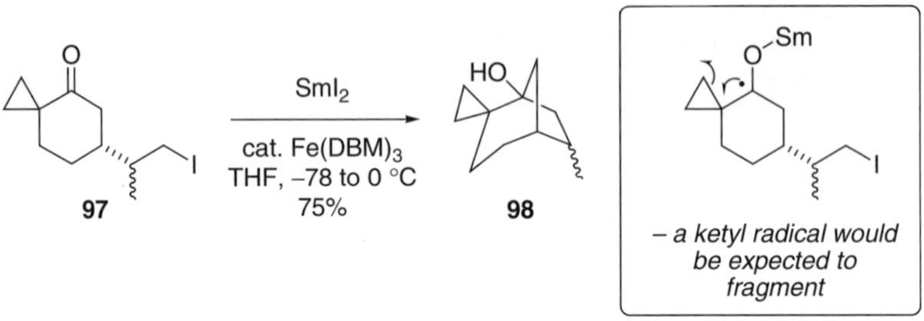

97 → 98: SmI_2, cat. $Fe(DBM)_3$, THF, –78 to 0 °C, 75%

– a ketyl radical would be expected to fragment

Scheme 5.61

With two reducible groups within the same molecule, the mechanism of the *intra*molecular SmI_2-mediated Barbier reaction is even less clear and more difficult to investigate. Again, Curran has been integral in evaluating mechanistic evidence from Kagan's group and others.[108]

In 1991, Molander produced evidence for the intermediacy of organosamarium species in intramolecular Barbier reactions.[113] Treatment of iodoketone **97** with SmI_2 gave the cyclisation product **98** in good yield, with no observed fragmentation of the cyclopropyl ring. To add weight to this result, a model substrate (lacking the iodide) was treated under the same conditions, furnishing fragmented product in 81% yield. This indicated that although the generation of ketyl radicals with subsequent cyclopropyl cleavage can occur, a mechanism not involving ketyl radicals must be in operation in the Barbier reaction (Scheme 5.61).[113]

Despite extensive studies, the true nature of the SmI_2-mediated Barbier reaction has yet to be irrefutably elucidated. In 1997, Curran carried out mechanistic investigations that illustrated the ambiguity still associated with the reaction mechanism. In order to study the rate of reduction of alkyl radicals to anions in intramolecular SmI_2-mediated Barbier reactions, Curran designed mechanistic probe **99** that would function as a 'radical clock'.[114] Treatment of **99** with SmI_2 was expected to produce both **100** (from organosamarium addition) and rearranged products, resulting from radical cyclisation on to the pendant alkene (Scheme 5.62). The rate constant for cyclisation of the radical on to the alkene was known from experiments with **99** and n-Bu_3SnH.

Scheme 5.62

Surprisingly, upon treatment with SmI_2, only tertiary alcohol **100** was obtained from the reaction. This finding appears to be at odds with the idea that SmI_2-mediated Barbier reactions proceed through alkyl radicals that are then reduced to organosamarium intermediates: the formation of an alkyl radical should result in at least some cyclisation (Scheme 5.62).[114]

In summary, it now appears likely that the SmI_2-mediated Barbier reaction does not proceed by one discrete mechanism. In addition to substrate structure, many variables, including solvent, additives and order of addition, may alter the reaction pathway. For simplicity and for alignment with the intermolecular Barbier and Grignard mechanisms, the intramolecular addition of organosamariums to carbonyl groups is the mechanistic path referred to throughout the remainder of this chapter. Readers should bear in mind, however, the alternative mechanisms that are outlined in simplified form in Scheme 5.63.

5.4.3 Intermolecular SmI_2-mediated Barbier and Grignard Reactions

5.4.3.1 Additions Involving Unfunctionalised Organohalides

In 1980, Kagan reported the first detailed study of the SmI_2-mediated Barbier reaction,[98] and showed the clear difference in behaviour of primary, secondary and tertiary alkyl halides in such couplings (Scheme 5.64). It is important to note that the reaction of *i*-BuMgBr or *s*-BuMgBr with ketones typically leads to reduction of the carbonyl group, so the use of SmI_2 to mediate such additions is noteworthy.

Representative procedure,[98] To a stirred solution of SmI_2 (2 mmol) in THF (50 ml) under nitrogen was added the carbonyl compound (1 mmol) in THF (5 ml), followed by the halide (1 mmol) in THF (5 ml). The reaction was carried out either at room temperature or in refluxing THF, with the reaction deemed complete when

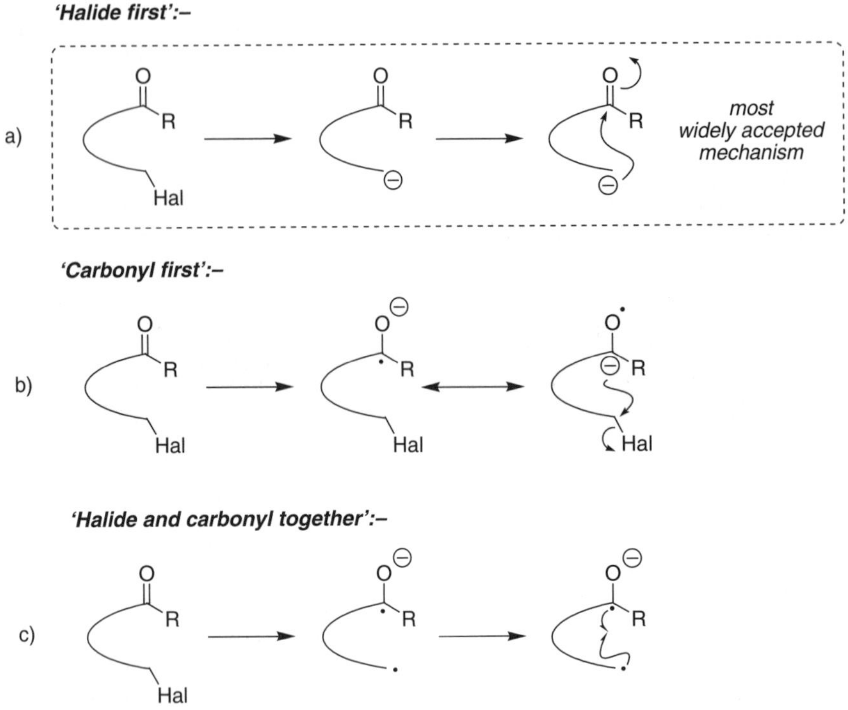

'Halide first':–

a) most widely accepted mechanism

'Carbonyl first':–

b)

'Halide and carbonyl together':–

c)

Scheme 5.63

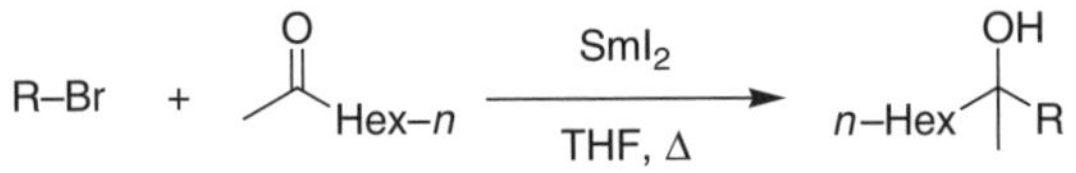

R–Br + (n-hexyl methyl ketone) —SmI_2, THF, Δ→ tertiary alcohol

R	reaction time	yield (%)
n–Bu	24 h	67%
s–Bu	36 h	27%[a]
i–Bu	48 h	33%[a]
t–Bu	96 h	9%[a]

[a] Recovered ketone predominated.

Scheme 5.64

the initial blue–green colour turned yellow. The reaction was worked up with 0.1 M HCl and extracted with Et_2O. The organic layers were washed with water, sodium thiosulfate, water and then brine and dried ($MgSO_4$). The solvent was removed *in vacuo* and the products were analysed by GLC.

Kagan also showed that allylic, benzylic and propargylic halides undergo efficient reaction with ketones (Scheme 5.65).[98] Importantly, in the absence of ketone, these alkyl halides react rapidly with SmI_2 to yield Wurtz homocoupled products, hence the use of Barbier conditions is crucial.

Kagan later found that THP was a better solvent than THF for the generation and reaction of allylic and benzylic samariums.[107] When THP is used, the allylation of aryl ketones (which are prone to pinacol coupling) is possible, provided that Grignard conditions are employed and ketones that are prone to enolisation are efficiently allylated under Barbier conditions (Scheme 5.66).[107]

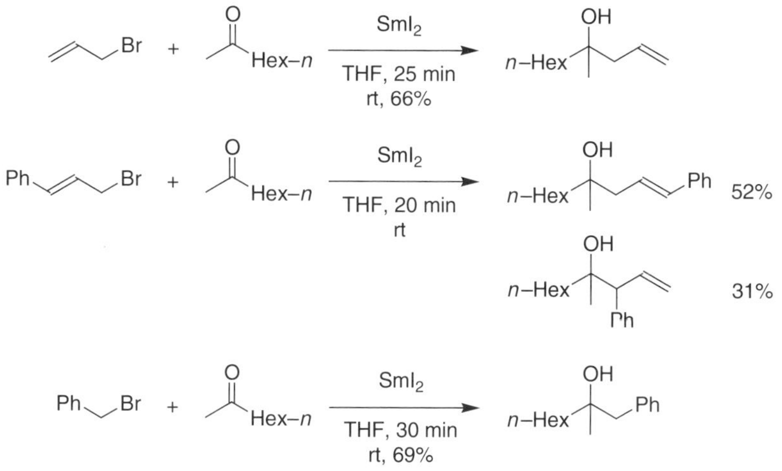
Br
+
O
Hex–n
SmI2
THF, 25 min
rt, 66%
OH
n–Hex
Ph
Br
+
O
Hex–n
SmI2
THF, 20 min
rt
OH
n–Hex
Ph
52%
OH
n–Hex
Ph
31%
Ph
Br
+
O
Hex–n
SmI2
THF, 30 min
rt, 69%
OH
n–Hex
Ph

Scheme 5.65

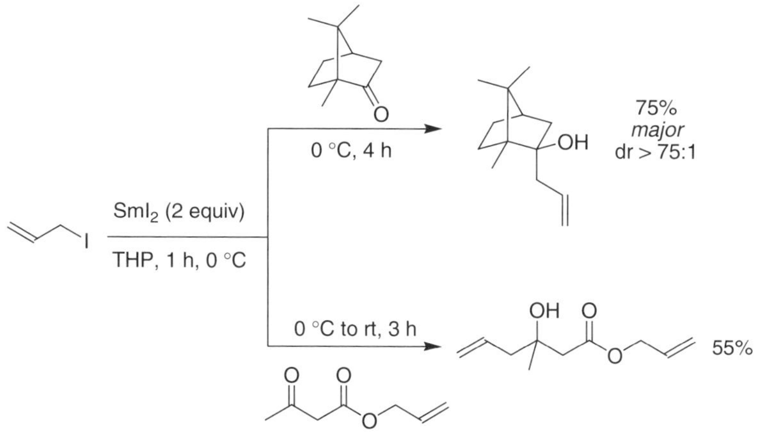
O
0 °C, 4 h
OH
75%
major
dr > 75:1
I
SmI2 (2 equiv)
THP, 1 h, 0 °C
0 °C to rt, 3 h
OH O
O
55%
O O
O

Scheme 5.66

Representative procedure.[107] To a stirred solution of SmI_2 (2 equiv) in THP was added the allyl iodide (1 equiv) at 0 °C and the mixture was stirred for 1 h. The substrate (0.45–0.90 equiv) was then added and the reaction was allowed to warm to room temperature. Once the reaction was complete, the mixture was subjected to acidic hydrolysis and the corresponding product isolated.

As discussed previously, vinyl, alkynyl and aryl halides are reduced by SmI_2 in THF to form radicals, but it is thought that the reaction does not proceed to form organosamariums due to competing radical processes. In some cases, the radical intermediates formed can translocate by hydrogen atom abstraction[115] or cyclisation,[112] to give new radicals. The radicals produced are then reduced by SmI_2 and the resulting organosamariums intercepted by carbonyl electrophiles (Scheme 5.67).

To prevent competing hydrogen atom abstraction from THF, Kunishima and Tani have shown that benzene can be used as the solvent. For example, iodobenzenes,[109] iodoalkynes[116,117] and vinyl halides[118] undergo efficient

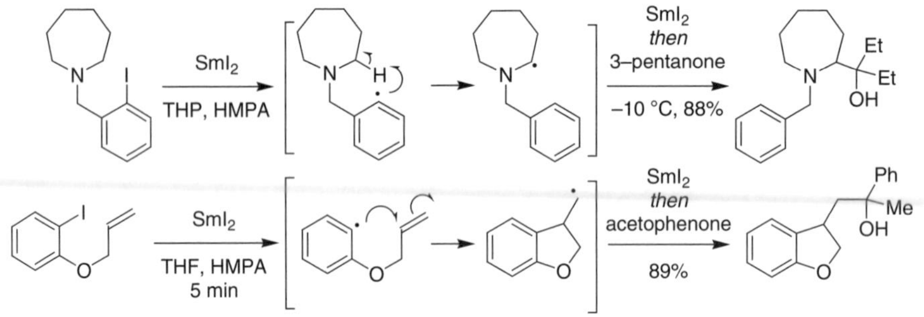

Scheme 5.67

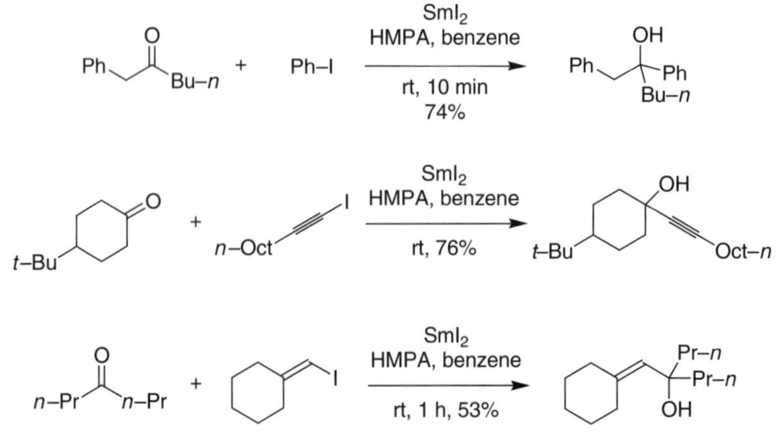

Scheme 5.68

addition to ketones and aldehydes upon treatment with SmI_2–HMPA in benzene (Scheme 5.68).

> *Representative procedure.* To a solution of SmI_2 (6 equiv) in benzene containing HMPA (10%) was added a mixture of the ketone (1 equiv) and organoiodide (3 equiv) and the reaction mixture was stirred at room temperature. Decolorisation of the resultant solution indicated complete reaction.

In some cases vinylsamariums have been generated and used in THF. For example, Kim reported the reduction of α-bromoacrylamides and the addition of the resulting vinylsamarium to aldehydes and ketones (Scheme 5.69).[119] The generation of vinylsamariums under these conditions may be possible due to an increased rate of reduction of the initially formed vinyl radical. Alternatively, the vinyl radical intermediate may abstract a hydrogen atom from THF, leading to a tetrahydrofuranylsamarium[117] that deprotonates the acrylamide to give a vinylsamarium.

5.4.3.2 Additions Involving Functionalised Organohalides

A variety of functionalised organic halides have been found to be compatible with the SmI_2-mediated Barbier and Grignard reactions. For example, benzyloxymethyl chloride can be used in conjunction with SmI_2 for the alkoxymethylation of ketones.[120] This addition has been found to work well even for hindered ketones such as **101**, an intermediate in White's 1987 approach to 2-desoxystemodinone (Scheme 5.70).[121] It is important to note that, in contrast to analogous organometallics, decomposition *via* α-elimination is not observed.

1,1-Dihaloalkanes have also been used in SmI_2-mediated Barbier reactions. For example, SmI_2 mediates the addition of diiodomethane to a range of aldehydes and ketones (Scheme 5.71).[122] In the case of additions to aldehyde **102**, high diastereocontrol is observed.

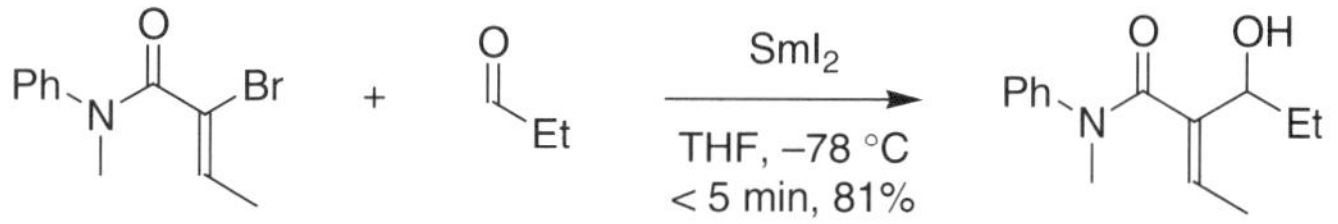

Ph N Br + O Et — SmI_2, THF, –78 °C, < 5 min, 81% → Ph N O OH Et

Scheme 5.69

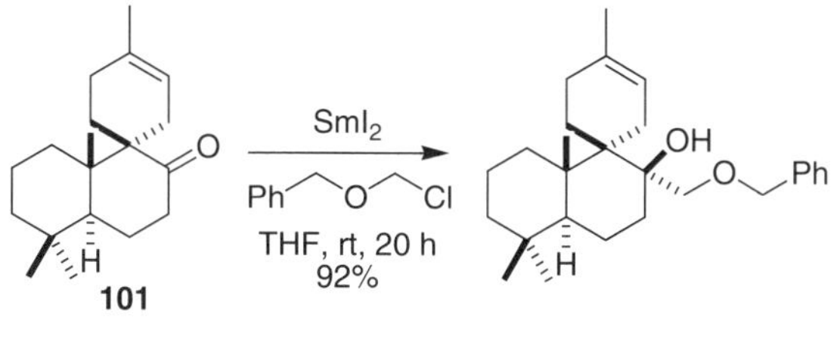

101 — SmI_2, Ph O Cl, THF, rt, 20 h, 92% → OH O Ph

Scheme 5.70

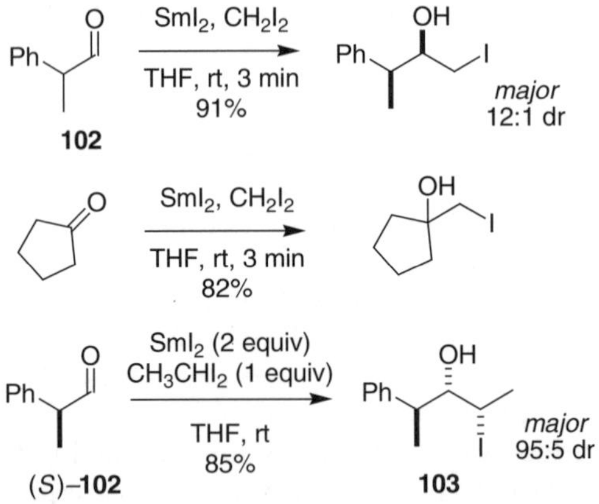

Scheme 5.71

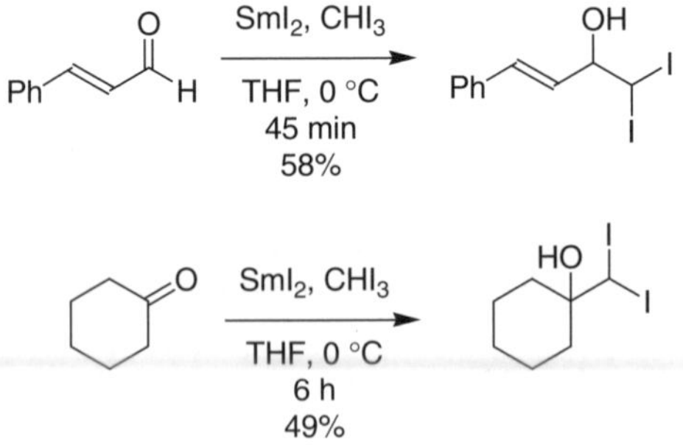

Scheme 5.72

Utimoto showed that the organosamarium formed from 1,1-diodoethane and SmI_2 underwent addition to aldehydes, such as (*S*)-**102**, to give **103** with near perfect diastereocontrol (Scheme 5.72).[123] In this case, it appears that the chiral organosamarium undergoes rapid equilibration, thus allowing reaction with the enantiomerically pure aldehyde (*S*)-**102** to effect a dynamic kinetic resolution.[123] Iodomethylations of ketones have also been carried out using mixtures of Sm and CH_2I_2.[120] SmI_2 can also be used to mediate the addition of iodoform to aldehydes and ketones to give products of diiodomethylation (Scheme 5.72).[124] Fluorinated alkyl halides have also been used in additions to aldehydes and ketones using SmI_2.[125]

SmI_2 has also been used to mediate Barbier-type reactions using substrates other than alkyl halides. For example, the groups of Sinaÿ, Beau and Skrydstrup have popularised the SmI_2-mediated inter- and intramolecular Barbier reactions of glycosyl sulfides with ketones and aldehydes, as a means by which *C*-glycosides can be prepared (see Chapter 7, Section 7.5). Huang described the diastereoselective SmI_2-mediated Barbier reactions of 2-pyridyl sulfides such as **104** with ketones, to give hydroxyalkylpyrrolidines **105** in moderate yield (Scheme 5.73).[126]

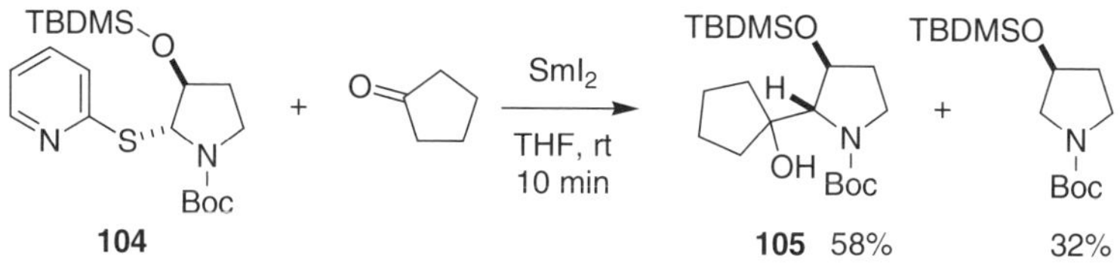

Scheme 5.73

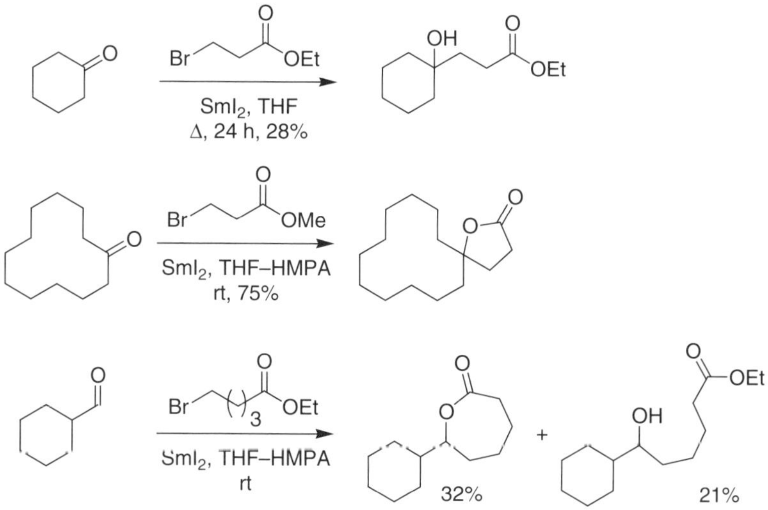

Scheme 5.74

Although organosamariums can react with esters and lactones, they preferentially undergo addition to aldehydes or ketones. This allows organosamariums bearing an ester functionality to be generated and used in intermolecular Barbier couplings (Scheme 5.74).[98,99]

Williams used SmI_2 to mediate the Barbier coupling of functionalised α-halo-2-methylazoles with aldehydes.[127] For example, coupling of iodide **106** with aldehyde **107** proceeded in good yield to give **108**, thus illustrating the value of the method for a future approach to phorboxazole A (Scheme 5.75).

Representative procedure.[127] To a stirred solution of SmI_2 (2.5 equiv) in anhydrous degassed THF was added via a cannula a mixture of the aldehyde **106** (1 equiv) and alkyl iodide **107** (1 equiv) dissolved in anhydrous degassed THF over 2 min. Once addition was complete, the reaction mixture was diluted with Et_2O and a saturated solution of Na/K tartrate and stirred until the organic phase was seen to separate from the resultant emulsion. The aqueous layer was extracted with hexanes–EtOAc (1:1) and the combined organic layers were washed with brine, dried using $MgSO_4$ and the solvent removed *in vacuo*. Crude **108** was purified by column chromatography (EtOAc–hexane eluent).

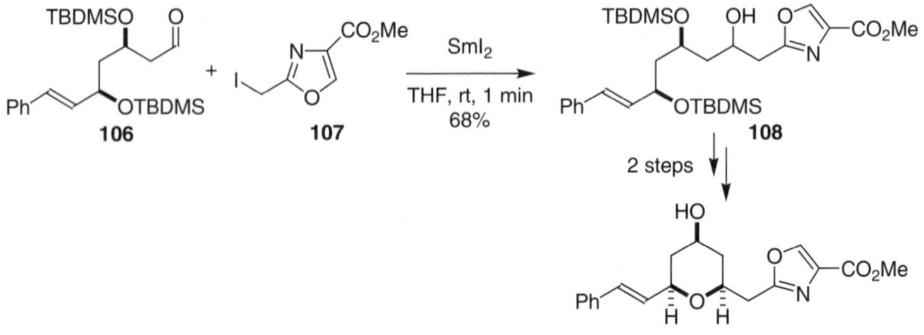

Scheme 5.75

5.4.4 Intramolecular SmI_2-mediated Barbier Reactions

SmI_2 is the reagent of choice for the cyclisation of a variety of halocarbonyl compounds.[97] A range of cyclic systems can be accessed using this method and cyclisations often occur with high levels of diastereocontrol. As discussed earlier, these reactions can be considered to proceed by reduction of the alkyl halide to the corresponding organosamarium and subsequent addition to the carbonyl group.

5.4.4.1 Intramolecular SmI_2-mediated Barbier Reactions Involving Addition to Aldehydes and Ketones

The SmI_2-mediated Barbier cyclisation of haloaldehydes and -ketones can be used for the formation of cyclic systems that typically range in size from three- to nine-membered rings. The formation of cyclobutanols using the method is unusual, but not unprecedented. Fukuzawa showed that treatment of bromoketones such as **109** (formed *in situ* by the addition of a Grignard reagent to the corresponding esters) with SmI_2 gave cyclic alcohols, including cyclopropanol **110**, in good yield.[128] Matsuda showed that medium rings can also be formed using the SmI_2 Barbier reaction: treatment of aldehydes bearing allylic chlorides with SmI_2 gave eight- and nine-membered cyclic alcohols in excellent yield (Scheme 5.76).[129]

Matsuda used the SmI_2 cyclisation of an aldehyde bearing an allylic chloride in a synthesis of the bicyclic skeleton of vinigrol (Scheme 5.77)[130] and subsequently showed that high-dilution conditions are not required and that the reaction is high yielding and general.[131]

A detailed study by Molander showed that the presence of ester and amide groups, α to the carbonyl group undergoing attack, in haloaldehyde and ketone substrates can lead to high levels of diastereocontrol in SmI_2-mediated Barbier cyclisations: the Lewis basic ester and amide carbonyls coordinate to samarium and control the stereochemical course of cyclisations (Scheme 5.78).[132]

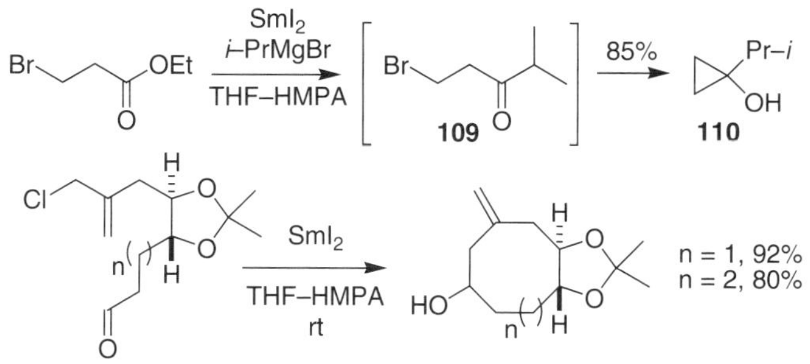

Scheme 5.76

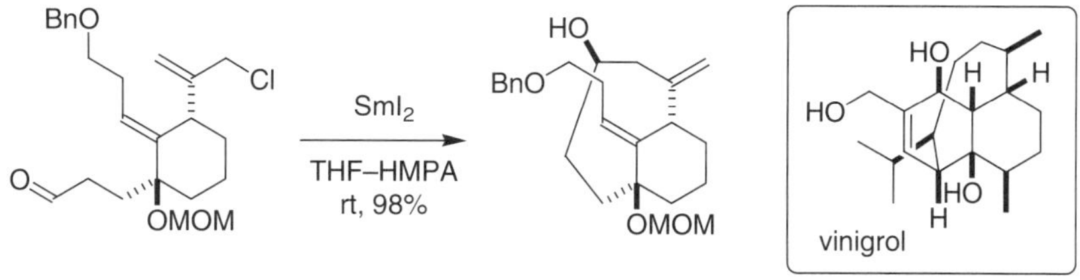

Scheme 5.77

Both alkyl halide and allylic halide groups undergo efficient Barbier cyclisation on treatment with SmI_2 between −78 °C and room temperature.[132]

The *syn* diastereoselectivity of the cyclisations most likely arises from kinetic control in which chelation of Sm(III) to the 1,3-dicarbonyl controls the orientation of the ketone prior to addition of the organosamarium (Scheme 5.79). However, thermodynamic control in which the diastereoisomeric products equilibrate by a retro-aldol–aldol sequence may operate in some cases.[132]

The SmI_2-mediated Barbier cyclisation of alkyl halides bearing cyclic ketones has been used extensively to form bicyclic systems.[103] When the alkyl halide tether is attached α to the cyclic ketone carbonyl, cyclisation typically proceeds to give *syn* products. For example, Cook used such a cyclisation in a two-directional strategy for the synthesis of tetracyclic polyquinines: treatment of bis-bromide **111** with SmI_2 in THF–HMPA gave diol **112** in 68% yield (Scheme 5.80).[133]

When the alkyl halide tether is not attached α to the cyclic ketone carbonyl group, SmI_2-mediated Barbier cyclisations result in bridged bicyclic systems. In 1991, Molander carried out a thorough survey of the use of SmI_2-mediated Barbier cyclisations to form bridged bicyclic systems and found that the Barbier reactions were best carried out at low temperature in the presence of catalytic $Fe(DBM)_3$ (Scheme 5.81).[113] Although these cyclisations are somewhat substrate dependent, there are many impressive examples in which SmI_2 can mediate the formation of strained ring systems and in which significant

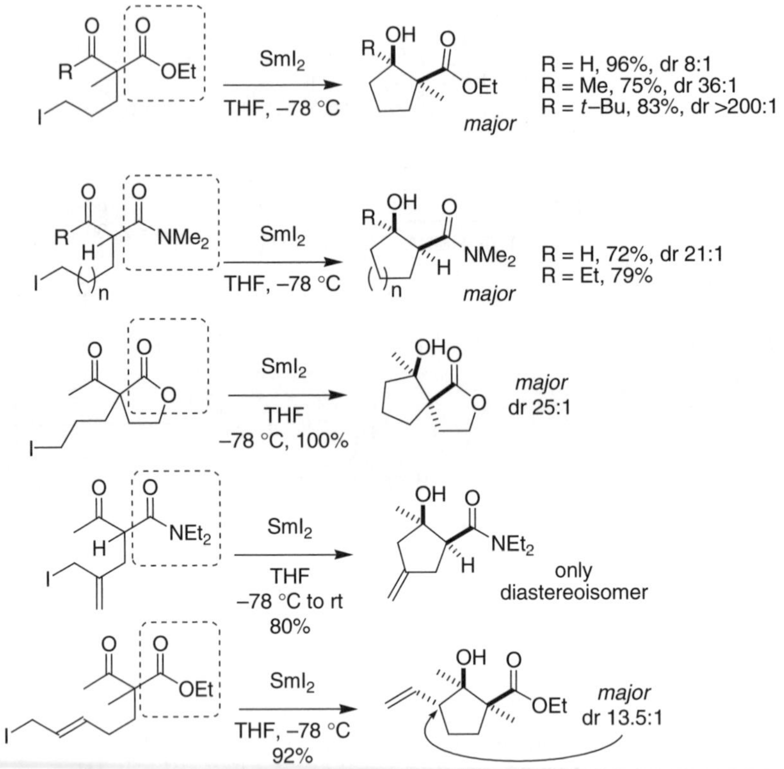

SmI_2, THF, –78 °C — *major*

R = H, 96%, dr 8:1
R = Me, 75%, dr 36:1
R = *t*–Bu, 83%, dr >200:1

SmI_2, THF, –78 °C — *major*

R = H, 72%, dr 21:1
R = Et, 79%

SmI_2, THF, –78 °C, 100% — *major* dr 25:1

SmI_2, THF, –78 °C to rt, 80% — only diastereoisomer

SmI_2, THF, –78 °C, 92% — *major* dr 13.5:1

Scheme 5.78

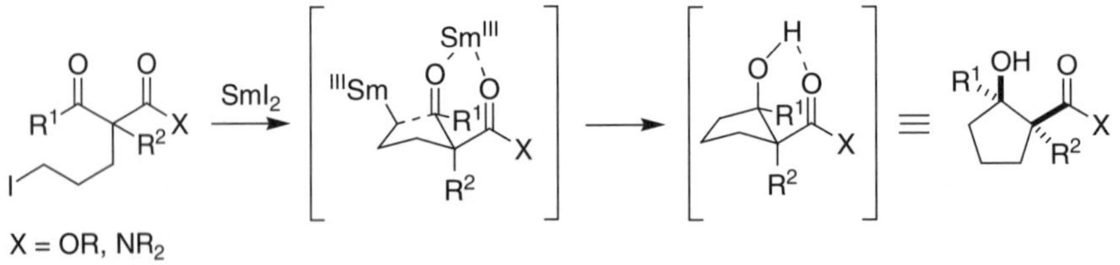

X = OR, NR_2

Scheme 5.79

111 → SmI_2, THF, HMPA, 68% → **112**

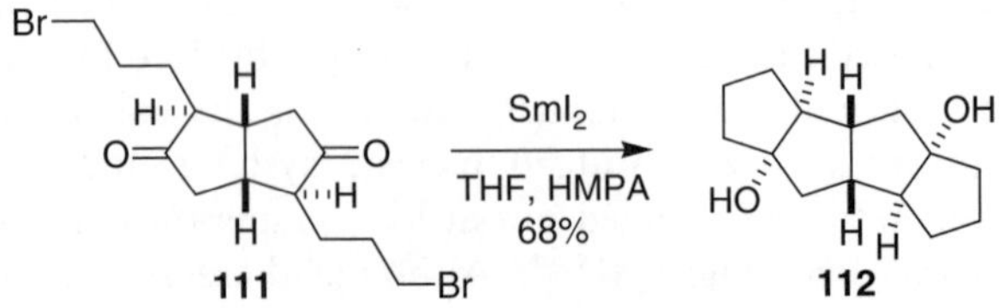

Scheme 5.80

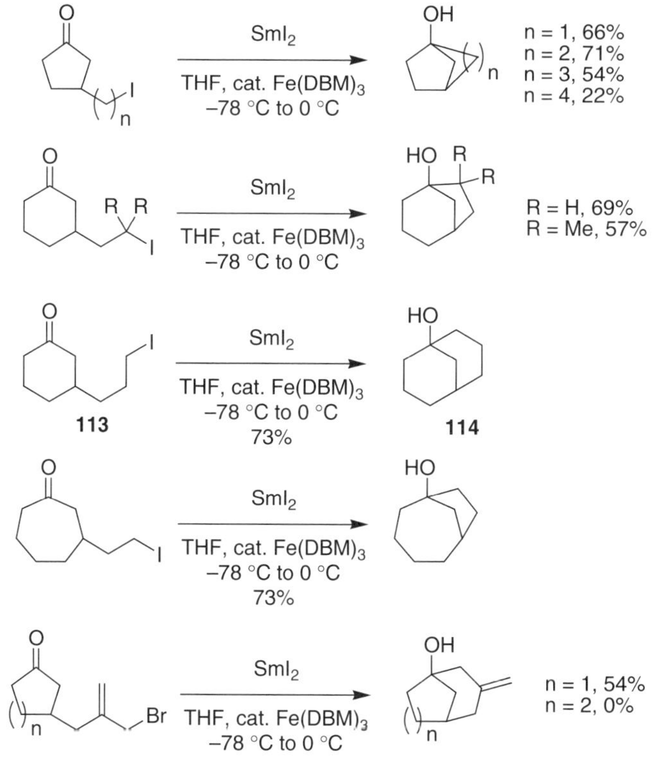

Scheme 5.81

steric congestion is overcome. It is interesting that the attempted cyclisation of iodide **113** using alternative conditions (*t*-BuLi, THF, −78 °C) did not give bicyclic alcohol **114**.[113] Finally, Molander showed that the more challenging cyclisation of substrates bearing allylic halide groups can give acceptable yields of bicyclic products (Scheme 5.81).[113]

Representative procedure.[113] To a stirred solution of SmI_2 (2 equiv) in THF was added $Fe(DBM)_3$ (1%) in THF and the reaction mixture was cooled to −78 °C. A pre-cooled solution of the haloketone (1 equiv) was then added using an ice-cooled cannula over 5 min and then the reaction mixture was allowed to warm to 0 °C. The reaction mixture was stirred for 2 h before being quenched with saturated $NaHCO_3$. The aqueous layer was extracted with Et_2O and the combined organic layers were washed with brine, dried ($MgSO_4$) and the solvent was removed. The crude product was purified by column chromatography (pentane–Et_2O eluent).

Molander has developed a SmI_2-mediated Barbier cyclisation–Grob fragmentation sequence for the synthesis of medium-sized carbocycles.[134] For example, treatment of mesylate **115** with SmI_2 and catalytic NiI_2 gave cyclodecenone **116** in excellent yield (Scheme 5.82).

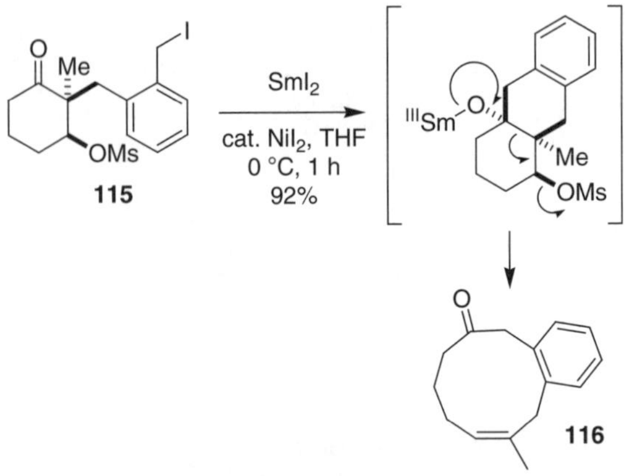

Scheme 5.82

5.4.4.2 *Intramolecular SmI_2-mediated Barbier Reactions Involving Addition to Esters and Amides*

In 1993, Molander found that in the presence of catalytic Fe(III) salts, SmI_2 mediates intramolecular Barbier additions to esters to give cyclic ketones (or cyclic hemiketals, if they prove to be stable).[135] Double addition to the ester is not observed, nor is reduction of the cyclic ketone product. This suggests that the tetrahedral intermediate, a samarium alkoxide of a cyclic hemiketal, is partially stable to the reaction conditions and the ketone group is not revealed until work-up. Molander found that both alkyl and allyl halides could be used in the additions (Scheme 5.83).[135]

As alkyl iodides and bromides are reduced faster than alkyl chlorides by SmI_2, it is possible to set up sequenced Barbier reactions involving dihalo ester substrates such as **117**:[135] under Molander's standard conditions [SmI_2, catalyst $Fe(DBM)_3$, low temperature], ketone **118** was obtained, whereas under more forcing conditions (SmI_2, tripiperidinophosphonamide), the alkyl chloride was also reduced and bicyclic alcohol **119** was formed in good yield (Scheme 5.84) (for further examples of sequential carbon–carbon bond-forming reactions using SmI_2, see Chapter 6).[135]

Molander used the SmI_2-mediated Barbier reaction of iodo esters to transfer the acyl group of the ester to carbon.[136] For example, treatment of iodo ester **120** with SmI_2 gives ketone **121** in 76% overall yield after *in situ* acetylation (Scheme 5.85).

Representative procedure.[136] To SmI_2 (2.4 equiv) in THF was added $Fe(DBM)_3$ (<0.001%) in THF by cannula and the solution was stirred for 1.5 h at room temperature. The resultant solution was cooled to −78 °C and then a solution of the iodo ester (1 equiv) in THF was added using a syringe pump. The reaction mixture was allowed to warm to room temperature over 30 min and then stirred for a further 1.5 h. The reaction was quenched with a saturated solution of $NaHCO_3$ and extracted with EtOAc. The combined organic layers were dried

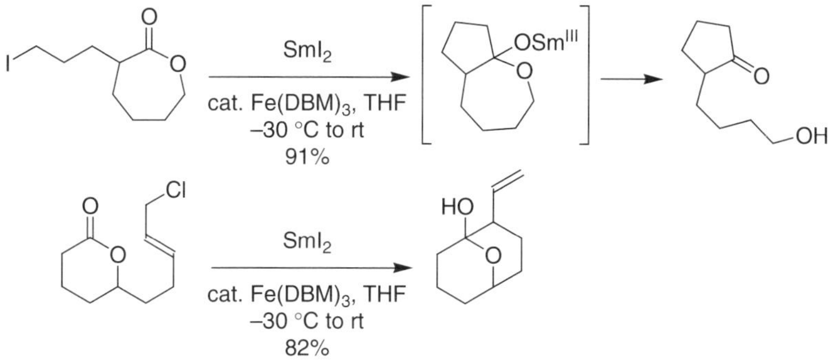

SmI_2
cat. $Fe(DBM)_3$, THF
–30 °C to rt
91%
OSm^{III}
OH

SmI_2
cat. $Fe(DBM)_3$, THF
–30 °C to rt
82%
HO

Scheme 5.83

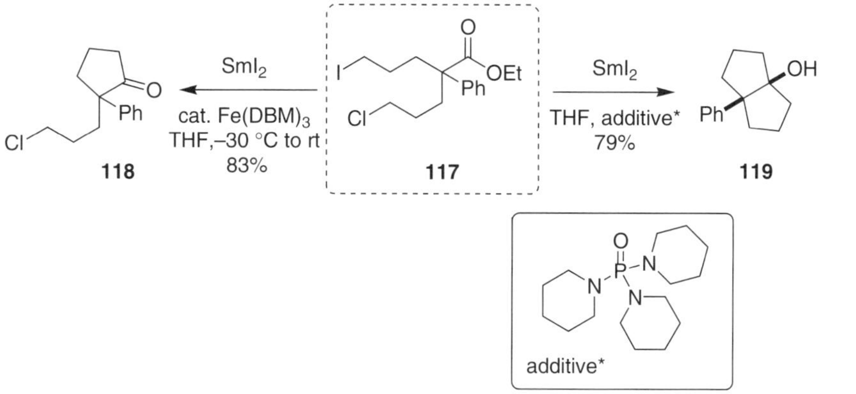

SmI_2
cat. $Fe(DBM)_3$
THF,–30 °C to rt
83%
118
OEt
Ph
117
SmI_2
THF, additive*
79%
OH
119
additive*

Scheme 5.84

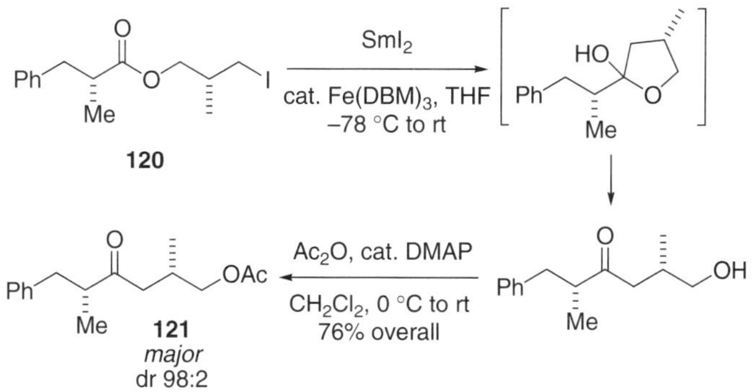

Ph
Me
120
SmI_2
cat. $Fe(DBM)_3$, THF
–78 °C to rt
HO
Ac$_2$O, cat. DMAP
CH_2Cl_2, 0 °C to rt
76% overall
OAc
121
major
dr 98:2
OH

Scheme 5.85

SmI$_2$ (2 equiv), THF–HMPA, −78 °C to −20 °C, 95%

SmI$_2$ (3 equiv), Fe(DBM)$_3$, THF, 0 °C; 35% combined

Scheme 5.86

($MgSO_4$) and the solvent was evaporated to yield the crude product, which was purified by column chromatography (EtOAc–hexanes eluent). This compound was acetylated without further purification.

The SmI$_2$-mediated intramolecular Barbier addition to amides is less common: Fadel reported the diastereoselective cyclisation of chloroimide **122**,[137] while Ha carried out an intramolecular Barbier addition to the carbonyl of cyclic imide **123** (Scheme 5.86).[138]

5.4.5 Conclusions

The coupling of organohalides with carbonyl groups is one of the most powerful transformations mediated by SmI$_2$. Although the mechanism of these carbon–carbon bond-forming reactions is not always clear, both intra- and intermolecular couplings can be carried out efficiently to access a range of products that can be difficult to obtain using more traditional methods. Molander recognised the potential of the SmI$_2$-mediated Barbier addition to esters for the initiation of sequential processes that assemble complex molecular architectures in one step. A discussion of these sequential carbon–carbon bond-forming processes can be found in Chapter 6.

5.5 SmI$_2$-mediated Reformatsky and Aldol-type Reactions

5.5.1 Introduction

Over the past 30 years, the widespread use of SmI$_2$ in organic synthesis has brought the chemistry of Sm(III) enolates to the fore as many processes using the reagent involve the formation and reaction of these organometallic species. Although Sm(III) enolates can be exploited as nucleophiles in a number of reactions, their reactivity is still poorly understood and they appear to behave differently to more conventional metal enolates; for example, there are few

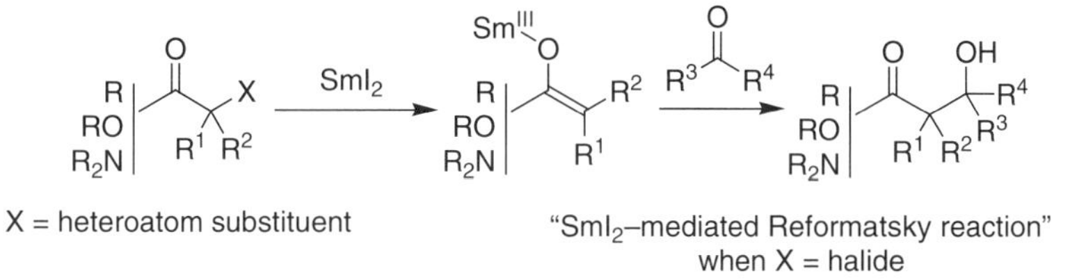

Scheme 5.87

examples of the alkylation of Sm(III) enolates with alkyl halides. Readers are directed to a recent review on the application of samarium enolates in organic synthesis for a more detailed discussion of these interesting organometallic intermediates.[139]

In Chapter 4, Section 4.5, we saw that SmI_2 reduces a range of α-heteroatom-substituted carbonyl compounds to the parent carbonyl compounds. These reductions proceed through intermediate Sm(III) enolates that are typically protonated by an alcohol cosolvent. The reduction of α-heteroatom-substituted carbonyl compounds is the most popular method for the generation of Sm(III) enolates. Although we are still some way from fully understanding the reactivity of Sm(III) enolates, they are known to undergo facile aldol reactions with aldehydes and ketones. When the heteroatom substituent in the starting material is a halide, these reactions are commonly referred to as SmI_2 Reformatsky reactions, but it is important to recognise that these reactions are a subset of a much larger class of SmI_2 aldol-type reactions (Scheme 5.87).

The SmI_2-mediated Reformatsky reaction provides a useful alternative to traditional versions of the reaction as it proceeds under mild, homogeneous conditions, with high chemo- and diastereoselectivity. Although α-halo esters are the most common substrates for the reaction, in principle any α-halo carbonyl compound can be employed in the reaction. The reactions are most often carried out by the addition of SmI_2 to a 1:1 mixture of the α-halocarbonyl compound and the coupling partner. These are often referred to as *Barbier conditions* (see Section 5.4).

5.5.2 Intermolecular Reformatsky Reactions with SmI_2

In 1980, Kagan reported the first example of a SmI_2-mediated Reformatsky reaction.[98] Kagan's study involved the coupling of ethyl α-bromopropionate with cyclohexanone (Scheme 5.88).

Analogous asymmetric, samarium Reformatsky reactions of chiral 3-bromoacetyl-2-oxazolidinones have been described by Fukuzawa.[140] For example, reduction of **124** with SmI_2 generates a samarium enolate that then reacts with pivalaldehyde to give the α-unbranched β-hydroxycarboximide **125** in 87% yield and in high diastereoisomeric excess (Scheme 5.89). The reaction is synthetically noteworthy as highly diastereoselective acetate aldol processes are difficult to achieve. Sm(III) ions are likely to play an important role in the

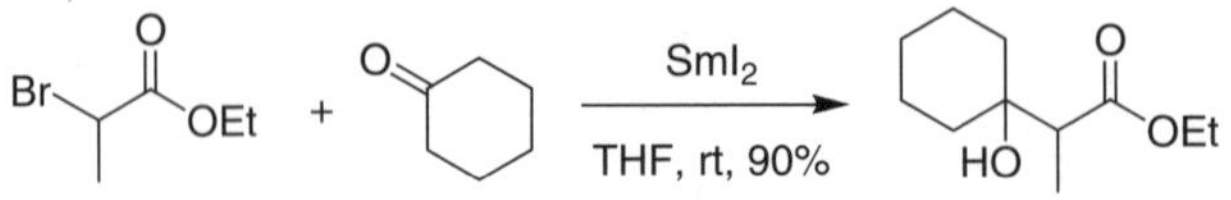

Scheme 5.88

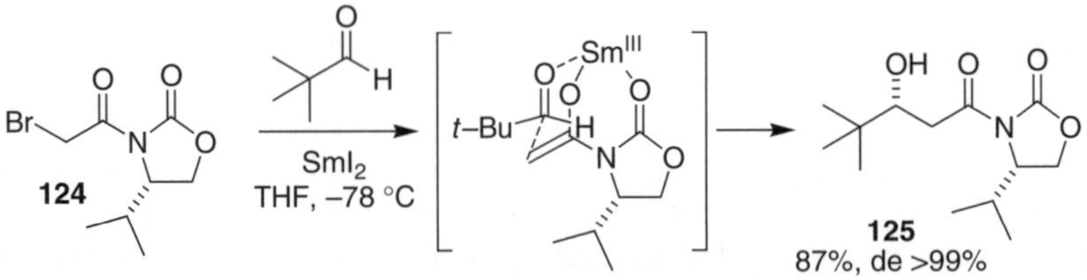

Scheme 5.89

transition state of the reaction leading to high diastereoselectivity.[140] As one would expect, coordination of the incoming electrophile to the Sm of the Sm(III) enolate has an important influence on reactivity and diastereoselectivity in all SmI_2-mediated Reformatsky and aldol-type reactions. Lewis acid activation of the electrophile may explain why Sm(III) enolate aldol-type reactions are relatively common whereas the reactions of Sm(III) -enolates with alkyl halides are scarce.

Representative procedure. To a solution of SmI_2 in THF (2.2 mmol) at –78 °C was added a mixture of the aldehyde (1.0 mmol) and acylated oxazolidinone (1.0 mmol) in THF dropwise over 5 min. The resulting solution was then stirred at –78 °C for 30 min, during which time the reaction mixture decolourised. The solution was then quenched with 0.1 M HCl and the aqueous layer extracted with Et_2O. The organic layers were washed with sodium thiosulfate and brine, dried ($MgSO_4$) and concentrated *in vacuo*. The crude product was purified by preparative TLC (ethyl acetate–hexane as eluent).

More recently, Concellón has reported a stereoselective method for the formation of (*E*)-α,β-unsaturated esters that exploits a SmI_2 Reformatsky reaction followed by an elimination.[141] For example, ethyl dibromoacetate reacts with benzaldehyde in the presence of SmI_2 to form samarium alkoxide **126**, which is reduced further to give a second Sm(III) enolate **127**. Elimination then affords (*E*)-α,β-unsaturated ester **128** in good yield (Scheme 5.90).[141]

In contrast, Concellón also showed that ethyl dibromoacetate undergoes a SmI_2 Reformatsky reaction with ketones to give 2-bromo-3-hydroxy esters as a mixture of diastereoisomers.[142] In a related study, Concellón showed that iodoacetic acid undergoes efficient SmI_2 Reformatsky reaction with aldehydes and ketones to give 3-hydroxycarboxylic acids directly.[143] Utimoto and Matsubara constructed samarium enolates, such as **129**, *in situ* from α-bromo esters using SmI_2 and found that they undergo efficient aldol reactions.[144] Quenching the samarium enolates with DCl in D_2O shows that the enolates are stable at

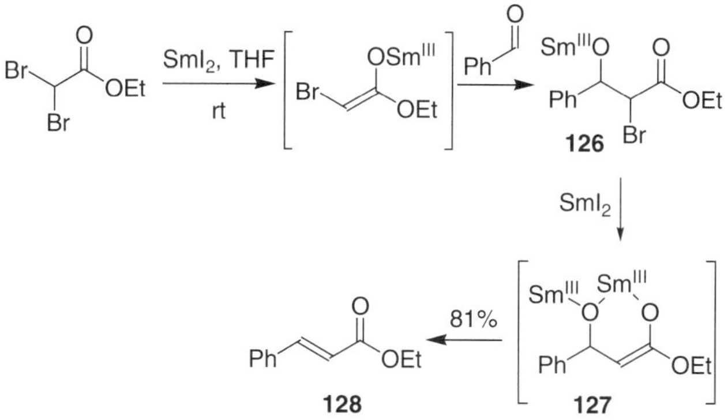

Scheme 5.90

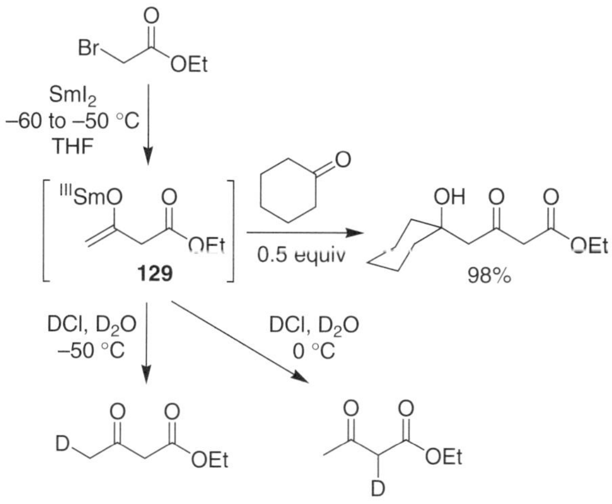

Scheme 5.91

$-50\ ^\circ C$ but isomerise to the more stable enolate on warming (Scheme 5.91). The use of two different α-halo esters allows access to more complex samarium enolates before quenching with benzaldehyde.[144]

A SmI_2-mediated Reformatsky reaction involving formaldehyde as the electrophile has been exploited by Otaka in a synthesis of a dipeptide isostere: treatment of γ,γ-difluoro-α,β-enoate **130** with SmI_2 in the presence of *in situ*-generated formaldehyde gave adduct **131** in good yield (Scheme 5.92).[145]

SmI_2 Reformatsky methodology is now sufficiently robust that Linhardt employed Reformatsky-type addition to carbon electrophiles in a solid-phase synthesis of *C*-sialosides (see Chapter 7, Section 7.3).[146] The use of α-halo ketones in intermolecular Reformatsky reactions is less common and is typically restricted to simple coupling partners. For example, Ohta showed that

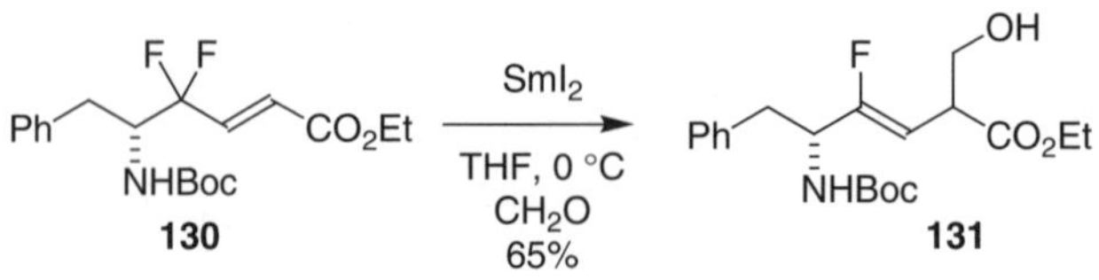

Scheme 5.92

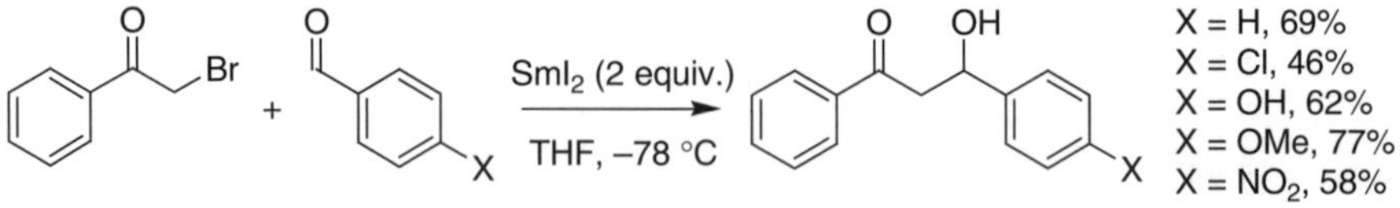

Scheme 5.93

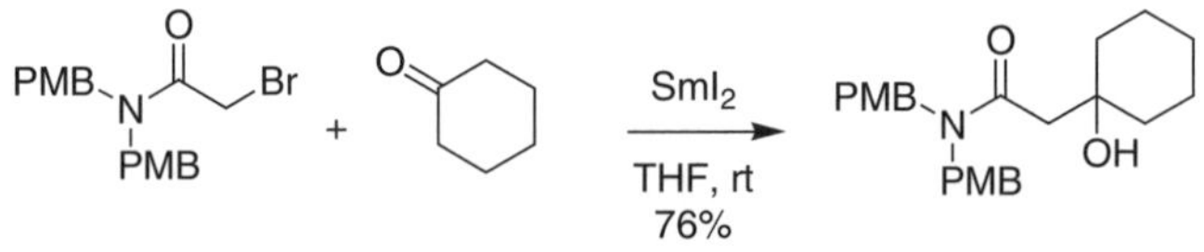

Scheme 5.94

phenacyl bromides undergo efficient coupling with benzaldehydes to give β-hydroxy ketones in moderate to good yield (Scheme 5.93).[147] The main limitation is presumably that the product ketones can undergo further reduction by SmI_2.

The intermolecular SmI_2 Reformatsky reactions of α-haloamides have similarly received little attention; however, Ohta showed that *N*,*N*-dibenzyl-α-haloamides undergo efficient coupling with a range of aldehydes and ketones (Scheme 5.94).[148]

5.5.3 Intramolecular Reformatsky Reactions with SmI_2

It is in the intramolecular sense that the SmI_2-mediated Reformatsky reaction has been used to greatest effect. Rather surprisingly, there are several examples of intramolecular SmI_2 Reformatsky reactions that employ protic cosolvents. In some cases, the alcohol cosolvent is essential for efficient Reformatsky cyclisation.

The reaction has been used to access medium and large carbocycles and lactones. In 1986, Inanaga reported the construction of medium- and large-ring lactones using SmI_2-mediated intramolecular Reformatsky reactions[149] and in 1991 developed a general synthesis of medium and large carbocycles by means of the Reformatsky reaction.[150] For example, the cyclisation of the Sm(III) enolate intermediates **132** gave carbocycles **133** in good yield. Inanaga suggested that the successful formation of such rings could be assisted by the large ionic radius of samarium, its flexible coordination number and its high oxophilicity, the lanthanide centre effectively bringing the two reacting centres together through chelation (Scheme 5.95).[150]

In 1991, Molander investigated the diastereoselectivity of SmI_2-mediated Reformatsky cyclisations and found that they often proceed with high levels of selectivity.[151] For example, treatment of α-bromo ester **134** with SmI_2 gave lactone **135** in 98% yield and as a single diastereoisomer (Scheme 5.96).

Representative procedure. To a solution of SmI_2 in THF (2.1 mmol) at −78 °C was added the α-bromo ester substrate (1.0 mmol) and the reaction mixture was stirred at −78 °C for 1 h. Aqueous saturated NH_4Cl (15 ml) and Et_2O (15 ml) were then added and the aqueous layer was separated and extracted with Et_2O (3 × 5 ml). The combined organic extracts were washed with brine (5 ml), dried ($MgSO_4$) and concentrated *in vacuo* and the crude product was purified by column chromatography.

Mukaiyama used SmI_2 Reformatsky cyclisations in a programme that culminated in an impressive total synthesis of paclitaxel (Taxol) (see Chapter 7, Section 7.4).[152]

n	*yield*
4	69%
5	70%
7	75%
11	82%

Scheme 5.95

Scheme 5.96

5.5.4 Intermolecular Aldol-type Reactions with SmI_2

There are a growing number of examples of SmI_2 aldol-type reactions involving α-heteroatom-substituted carbonyl compounds where the group expelled from the starting material is not a halogen: oxygen-, nitrogen-, sulfur- and selenium-containing substrates have all found application in SmI_2-mediated aldol-type reactions.

In 1993, Enholm described the SmI_2-mediated aldol reactions of α-benzoyl lactones derived from carbohydrates with ketones.[153] For example, treatment of lactone **136** with SmI_2 in the presence of (+)-dihydrocarvone **137** gave aldol adduct **138** in good yield and in high diastereoisomeric excess (Scheme 5.97). In this example, HMPA is used as an additive to increase the reduction potential of SmI_2 and thus to facilitate Sm(III) enolate formation from the α-benzoyl lactone.[153]

Representative procedure. To a solution of SmI_2 in THF (5 mmol) and HMPA (300 µl) at −78 °C was added the lactone substrate (1 mmol) and the terpene ketone (2 mmol) in THF and the reaction mixture was warmed to room temperature. When TLC showed that the lactone starting material had been consumed, aqueous saturated $NaHCO_3$ and ethyl acetate were added and the resulting mixture stirred for 30 min. After filtration through Celite and concentration *in vacuo*, the crude product was purified by column chromatography.

Procter used the cleavage of an aryloxy substituent with SmI_2 to initiate aldol-type reactions.[154] For example, treatment of ketone **139** with SmI_2 in the presence of cyclohexanone and tetrahydropyran-4-one resulted in efficient aldol reaction to give adducts **140** and **141**, respectively (Scheme 5.98). Again, an

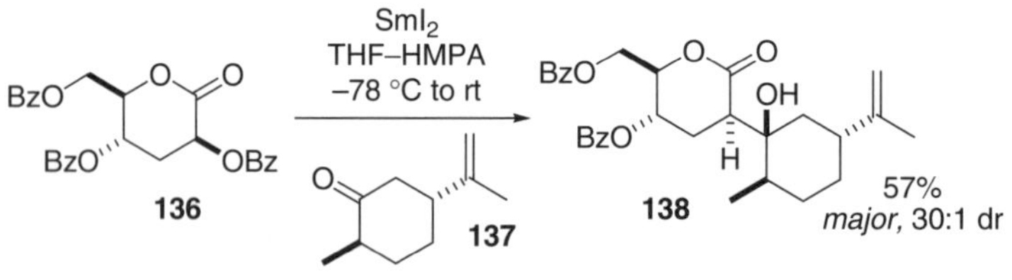

Scheme 5.97

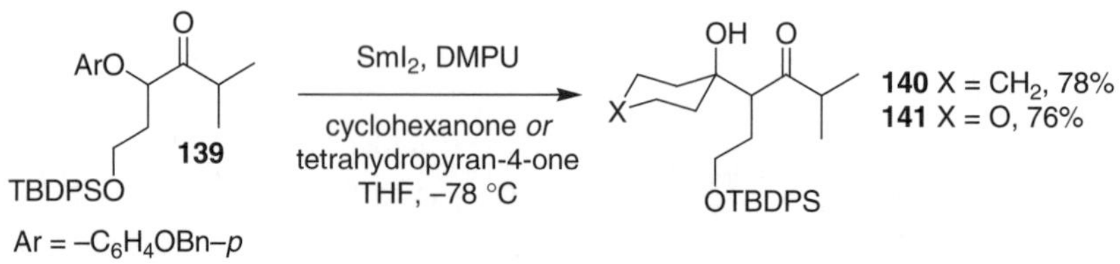

Scheme 5.98

additive is used, this time DMPU, to increase the reduction potential of SmI_2 and to facilitate Sm(III) enolate formation.[154]

In 1999, Matsuda utilised an intermolecular SmI_2-mediated aldol reaction, triggered by the reductive removal of an α-alkylsulfanyl group, in the first synthesis of herbicidin B.[155] The Sm(III) enolate **143** was generated by the reduction of glycosyl sulfide **142** with SmI_2. When TLC showed the reduction of **142** to be complete, oxygen was passed through the reaction mixture to destroy excess SmI_2 before the addition of aldehyde **144**. Aldol adduct **145** was obtained in high yield and as a mixture of diastereoisomers. This example is particularly noteworthy as Barbier conditions – the use of an *in situ* electrophilic quench – were not employed (Scheme 5.99).[155]

Shuto and Matsuda also utilised a SmI_2-mediated aldol reaction triggered by the reductive removal of an arylselenide group to construct 1′α-branched uridine derivatives.[156]

The chemoselectivity of SmI_2 and the low basicity of the resultant Sm(III) enolate species make the lanthanide reagent ideal for sensitive substrates. In 2000, Skrydstrup utilised a SmI_2-mediated aldol reaction to introduce carbinol side chains selectively into glycine residues in peptides.[157] For example, treatment of α-pyridyl sulfide tripeptide **146** with SmI_2 at room temperature gave Sm(III) enolate **147** that underwent aldol reaction with cyclohexanone to give modified peptide **148** in good yield (Scheme 5.100).[157]

SmI_2-mediated aldol reactions can also be carried out using epoxides and aziridines as precursors to the Sm(III) enolate. In 2000, Mukaiyama reported a method for the synthesis of unsymmetrical bis-aldols using SmI_2 to

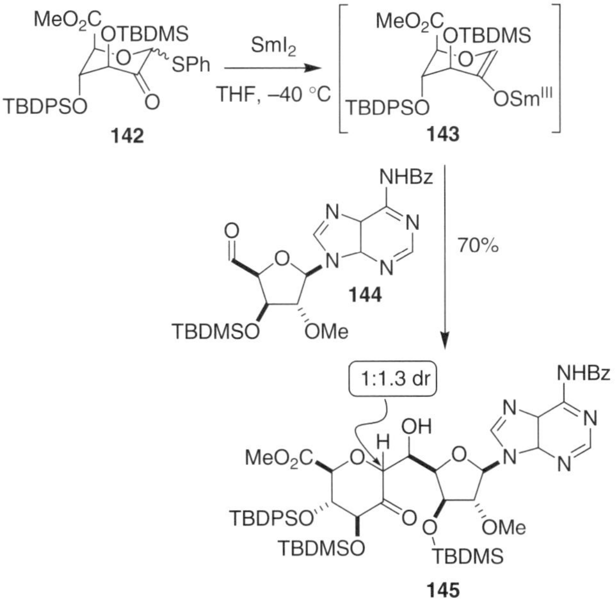

Scheme 5.99

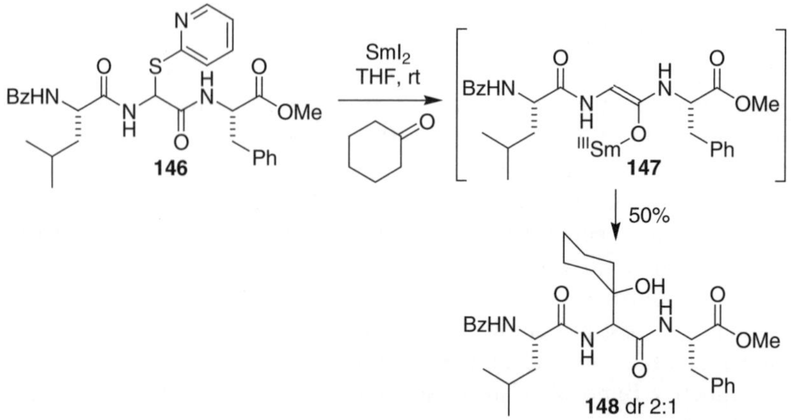

Scheme 5.100

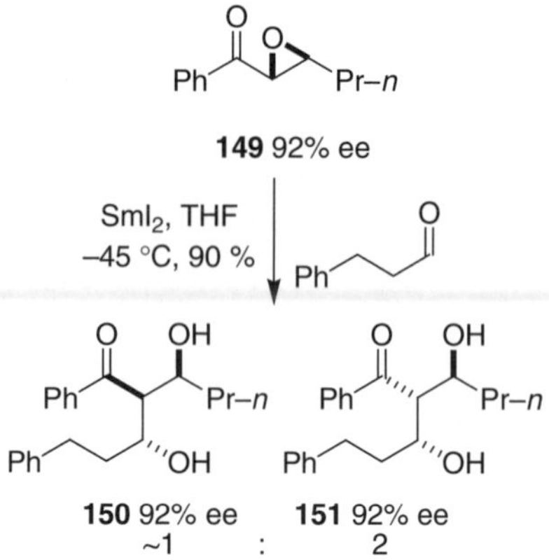

Scheme 5.101

mediate aldol reactions between aldehydes and aryl or alkyl oxiranyl ketones (Scheme 5.101).[158] Only the *syn,syn*-**150** and *anti,anti*-**151** bis-aldols were obtained from the reaction of epoxide **149**. Mukaiyama also reported an intramolecular variant of this aldol process.[159]

In 2005, Mukaiyama investigated the synthesis of β-amino-β′-hydroxy ketones using SmI_2-mediated aldol reactions between aldehydes and Sm(III) enolates derived from aziridinyl ketones.[160] Mukaiyama also examined an asymmetric SmI_2 aldol process using an unsaturated aziridine substrate bearing an Evans oxazolidinone auxiliary. For example, on treatment with SmI_2, aziridine **152** underwent smooth ring opening, Sm(III) enolate formation and a highly *syn* diastereoselective aldol process to give diastereoisomers **153** and **154** in good yield (Scheme 5.102).[161]

Scheme 5.102

5.5.5 Intramolecular Aldol-type Reactions with SmI_2

Intramolecular SmI_2 aldol-type reactions involving α-heteroatom-substituted carbonyl compounds where the group expelled from the starting material is not a halogen have also been carried out.

In 2002, Skrydstrup reported the diastereoselective construction of functionalised prolines using a SmI_2-mediated aldol cyclisation.[162] Treatment of β-lactam-derived α-benzoyloxy esters, such as **155**, with SmI_2 led to the generation of a Sm(III) enolate **156**, aldol cyclisation and addition of the resultant samarium alkoxide to the β-lactam carbonyl. The efficient sequential reaction gave proline derivatives, such as **157**, with high diastereoselectivity and in good yield (Scheme 5.103).[162] This example illustrates how the presence of a protic cosolvent does not necessarily interfere with the intramolecular aldol reaction and can in fact be crucial to the success of the cyclisation.

5.5.6 SmI_2 Aldol-type Reactions as Part of Sequential Processes

Although the reduction of α-heteroatom-substituted carbonyl compounds is the most popular method for the generation of Sm(III) enolates, other substrates react with SmI_2 to form Sm(III) enolates that undergo aldol-type reactions as part of larger, sequential processes.

The conjugate reduction of α,β-unsaturated carbonyl compounds is a potentially useful way of accessing Sm(III) enolates that has yet to be widely exploited. Cabrera reported a cyclodimerisation sequence of α,β-unsaturated ketones using SmI_2 that involves a diastereoselective aldol cyclisation.[163] For example, treatment of chalcone **158** with SmI_2 generates cyclopentanol **160** in quantitative yield after chelation-controlled aldol cyclisation of Sm(III) enolate **159** (Scheme 5.104).

Fang used similar reductive cyclisations of 1,1′-dicinnamoylferrocenes such as **161** to prepare 3-ferrocenophane diols **162**.[164] In this case, the aldol cyclisation is followed by ketone reduction in a highly diastereoselective, sequential operation (Scheme 5.105).

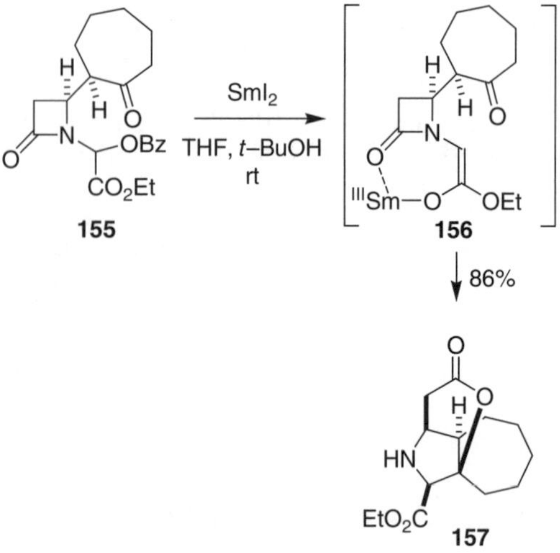

Scheme 5.103

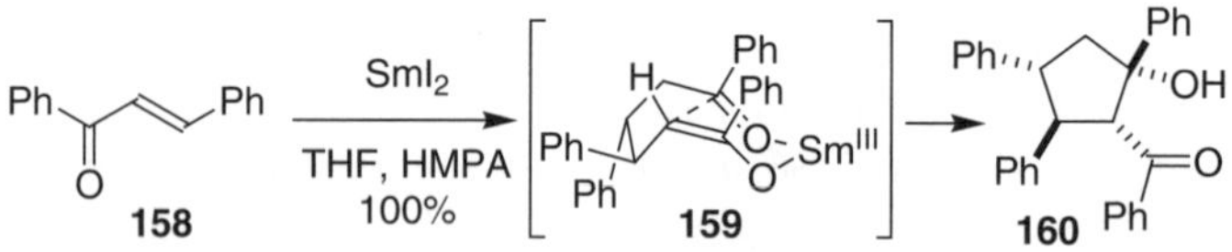

Scheme 5.104

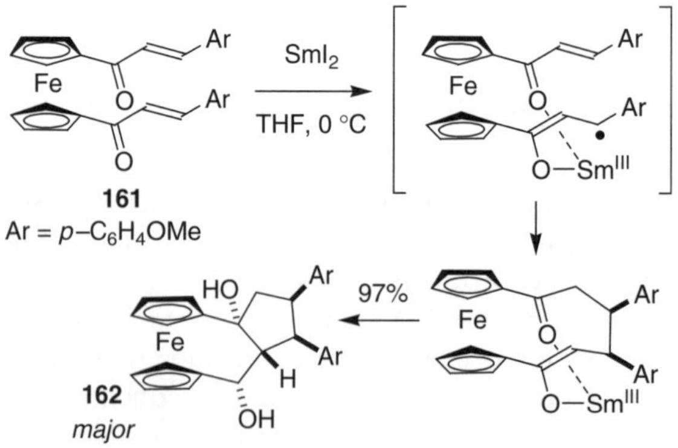

Scheme 5.105

Procter reported the diastereoselective spirocyclisation of unsaturated ketones **163** using SmI_2.[43,165] The cyclisation proceeds by conjugate reduction, Sm(III) enolate generation and chelation-controlled aldol cyclisation to give *syn*-spirocyclic cyclopentanols **164** in good yield (Scheme 5.106). It is important

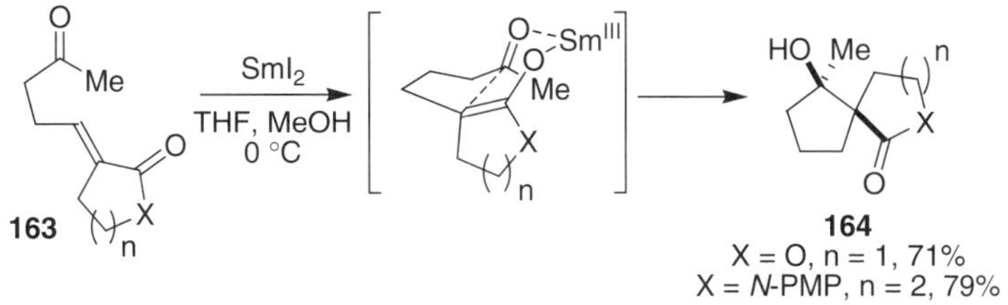

Scheme 5.106

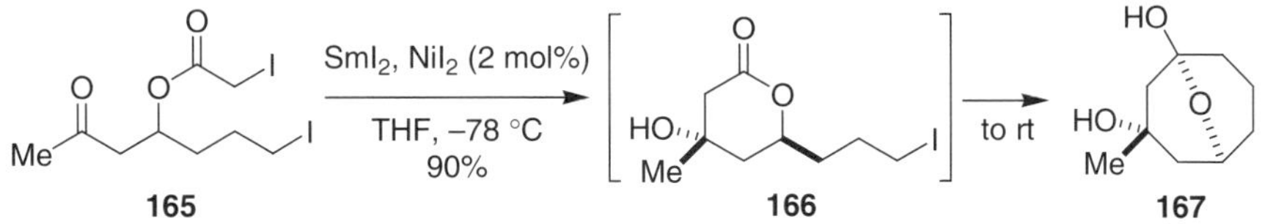

Scheme 5.107

to note that this reductive aldol sequence is only possible because the shortness of the tether between the ketone carbonyl and the enoate prohibits a carbonyl–alkene cyclisation pathway from operating. Again, the protic cosolvent is important for the success of the reductive aldol cyclisation presumably as the alcohol cosolvent promotes the conjugate reduction step.[43,165]

Representative procedure. To a solution of SmI_2 (0.1 M in THF, 2.80 mmol) at 0 °C was added MeOH (21.0 mmol) and the mixture was left to stir for 30 min. A solution of keto-lactam **163** (X = *N*-PMP) (0.70 mmol) in THF was then added by cannula and the reaction mixture was left to stir for 5 h. The reaction was quenched by opening to air and the subsequent addition of aqueous saturated NaCl. The aqueous layer was separated and extracted with EtOAc. The combined organic extracts were then dried ($NaSO_4$) and concentrated *in vacuo*. The crude product was purified by column chromatography (60% ethyl acetate–petroleum ether as eluent) to give **164** (X = *N*-PM P) (0.55mmol, 79%) as a colourless oil.

SmI_2 carbonyl–alkene couplings in which an α, β-unsaturated carbonyl compound is involved also generate Sm(III) enolates (see Section 5.2). In 1994, Enholm described carbonyl–alkene cyclisation–intermolecular aldol sequence using the reagent (see Chapter 6).[166] Finally, Molander developed a sequential SmI_2-mediated Reformatsky, nucleophilic acyl substitution reaction for the synthesis of medium-sized carbocycles.[167] Treatment of α-iodo ester **165** with SmI_2 generates iodolactone **166**, which then undergoes cyclisation to give **167** as a single diastereoisomer (Scheme 5.107). The NiI_2 catalyst present in the reaction mixture promotes reduction of the intermediate alkyl iodide to the corresponding alkylsamarium required for nucleophilic addition to the lactone carbonyl.[167] Further examples of SmI_2-mediated carbon–carbon bond-forming sequences can be found in Chapter 6.

5.5.7 Conclusions

Sm(III) enolates are important intermediates in many SmI_2 reactions. The reduction of α-halo-substituted carbonyl compounds with SmI_2 is the most popular method for the generation of Sm(III) enolates, although substrates bearing a range of other α-heteroatom substituents can be employed in SmI_2-mediated aldol-type reactions. Reformatsky and aldol reactions mediated by SmI_2 can be used in inter- and intramolecular fashion and offer considerable advantages over other metal-mediated reactions due to the high chemo-, regio- and diastereoselectivity of the reactions. The aldol reactions of Sm(III) enolates are becoming increasingly important as components of SmI_2-mediated cascade reactions.

References

1. R. Fittig, *Liebigs Ann. Chem.*, 1859, **110**, 23.
2. B. E. Kahn and R. D. Rieke, *Chem. Rev.*, 1988, **88**, 733.
3. J. E. McMurry, *Chem. Rev.*, 1989, **89**, 1513.
4. T. Wirth, *Angew. Chem. Int. Ed.*, 1996, **35**, 61.
5. A. Gansäuer and H. Bluhm, *Chem. Rev.*, 2000, **100**, 2771.
6. A. Chattergee and N. N. Joshi, *Tetrahedron*, 2006, **62**, 12137.
7. J.-L. Namy, J. Souppe and H. B. Kagan, *Tetrahedron Lett.*, 1983, **24**, 765.
8. A. Fürstner, R. Csuk, C. Rohrer and H. Weidmann, *J. Chem. Soc., Perkin Trans. 1*, 1988, 1729.
9. N. Akane, Y. Kanagawa, Y. Nishiyama and Y. Ishii, *Chem. Lett.*, 1992, 2431.
10. J. Inanaga, M. Ishikawa and M. Yamaguchi, *Chem. Lett.*, 1987, 1485.
11. T. B. Christensen, D. Riber, K. Daasbjerg and T. Skrydstrup, *Chem. Commun.*, 1999, 2051.
12. H. L. Pedersen, T. B. Christensen, R. J. Enemærke, K. Daasbjerg and T. Skrydstrup, *Eur. J. Org. Chem.*, 1999, 565.
13. N. Taniguchi, N. Kaneta and M. Uemura, *J. Org. Chem.*, 1996, **61**, 6088.
14. R. Nomura, T. Matsuno and T. Endo, *J. Am. Chem. Soc.*, 1996, **118**, 11666.
15. H. C. Aspinall, N. Greeves and C. Valla, *Org. Lett.*, 2005, **7**, 1919.
16. M. Adinolfi, G. Barone, A. Iadonisi, L. Mangoni and R. Manna, *Tetrahedron*, 1997, **53**, 11767.
17. G. A. Molander and C. Kenny, *J. Org. Chem.*, 1988, **53**, 2132.
18. G. A. Molander and C. Kenny, *J. Am. Chem. Soc.*, 1989, **111**, 8236.
19. S. Arseniyadis, D. V. Yashunsky, R. Pereira de Frietas, M. Muñoz Dorado, E. Toromanoff and P. Potier, *Tetrahedron Lett.*, 1993, **34**, 1137.
20. M. Carpintero, C. Jaramillo and A. Fernández-Mayoralas, *Eur. J. Org. Chem.*, 2000, 1285.
21. A. Kornienko, G. Marnera and M. d'Alarcao, *Carbohydr. Res.*, 1998, **310**, 141.

22. C. Anies, A. Pancrazi, J.-Y. Lallemand and T. Prangé, *Bull. Soc. Chim. Fr.*, 1997, **134**, 203.
23. T. Kan, S. Hosokawa, S. Nara, M. Oikawa, S. Ito, F. Matsuda and H. Shirahama, *J. Org. Chem.*, 1994, **59**, 5532.
24. C. S. Swindell and W. Fan, *Tetrahedron Lett.*, 1996, **37**, 2321.
25. J. L. Chiara, J. Marco-Contelles, N. Khiar, P. Gallego, C. Destabel and M. Bemabe, *J. Org. Chem.*, 1995, **60**, 6010.
26. C. F. Sturino and A. G. Fallis, *J. Am. Chem. Soc.*, 1994, **116**, 7447.
27. H. Miyabe, S. Kanehira, K. Kume, H. Kandori and T. Naito, *Tetrahedron*, 1998, **54**, 5883.
28. I. S. de Gracia, S. Bobo, M. D. Martin-Ortega and J. L. Chiara, *Org. Lett.*, 1999, **1**, 1705.
29. K. C. Nicolaou, X. Huang, N. Giuseppone, P. B. Rao, M. Bella, M. V. Reddy and S. A. Snyder, *Angew. Chem. Int. Ed.*, 2001, **40**, 4705.
30. K. C. Nicolaou, S. A. Snyder, N. Giuseppone, X. Huang, M. Bella, M. V. Reddy, P. B. Rao, A. E. Koumbis, P. Giannakakou and A. O'Brate, *J. Am. Chem. Soc.*, 2004, **126**, 10174.
31. H. Miyabe, M. Torieda, K. Inoue, K. Tajiri, T. Kiguchi and T. Naito, *J. Org. Chem.*, 1998, **63**, 4397.
32. D. Riber, R. Hazell and T. Skrydstrup, *J. Org. Chem.*, 2000, **65**, 5382.
33. G. Masson, S. Py and Y. Vallée, *Angew. Chem. Int. Ed.*, 2002, **41**, 1772.
34. O. N. Burchak, C. Philouze, P. Y. Chavant and S. Py, *Org. Lett.*, 2008, **10**, 3021.
35. Y. Tanaka, N. Taniguchi, T. Kimura and M. Uemura, *J. Org. Chem.*, 2002, **67**, 9227.
36. Y.-W. Zhong, Y.-Z. Dong, K. Fang, K. Izumi, M.-H. Xu and G.-Q. Lin, *J. Am. Chem. Soc.*, 2005, **127**, 11956.
37. E. J. Enholm, D. C. Forbes and D. P. Holub, *Synth. Commun.*, 1990, **20**, 981.
38. T. Imamoto and S. Nishimura, *Chem. Lett.*, 1990, 1141.
39. Y.-W. Zhong, K. Izumi, M.-H. Xu and G.-Q. Lin, *Org. Lett.*, 2004, **6**, 4747.
40. Y.-W. Zhong, M.-H. Xu and G.-Q. Lin, *Org. Lett.*, 2004, **6**, 3953.
41. J.-P. Ebran, R. G. Hazell and T. Skrydstrup, *Chem. Commun.*, 2005, 5402.
42. D. J. Johnston, E. Couché, D. J. Edmonds, K. W. Muir and D. J. Procter, *Org. Biomol. Chem.*, 2003, **1**, 328.
43. T. K. Hutton, K. W. Muir and D. J. Procter, *Org. Lett.*, 2003, **5**, 4811.
44. S. Fukuzawa, A. Nakanishi, T. Fujinami and S. Sakai, *J. Chem. Soc., Chem. Commun.*, 1986, 624.
45. S. Fukuzawa, A. Nakanishi, T. Fujinami and S. Sakai, *J. Chem. Soc., Perkin Trans. 1*, 1988, 1669.
46. K. Otsubo, J. Inanaga and M. Yamaguchi, *Tetrahedron Lett.*, 1986, **27**, 5763.
47. M. Kawatsura, F. Matsuda and H. Shirahama, *J. Org. Chem.*, 1994, **59**, 6900.

48. S. Fukuzawa, K. Seki, M. Tatsuzawa and K. Mutoh, *J. Am. Chem. Soc.*, 1997, **119**, 1482.
49. N. J. Kerrigan, P. C. Hutchison, T. D. Heightman and D. J. Procter, *Chem. Commun.*, 2003, 1402.
50. N. J. Kerrigan, P. C. Hutchison, T. D. Heightman and D. J. Procter, *Org. Biomol. Chem.*, 2004, 2476.
51. L.-L. Huang, M.-H. Xu and G.-Q. Lin, *J. Org. Chem.*, 2005, **70**, 529.
52. C. A. Merlic and J. C. Walsh, *J. Org. Chem.*, 2001, **66**, 2265.
53. G. A. Molander and J. A. McKie, *J. Org. Chem.*, 1995, **60**, 872.
54. G. A. Molander and J. A. McKie, *J. Org. Chem.*, 1992, **57**, 3132.
55. S. C. Shim, J.-T. Hwang, H.-Y. Kang and M. H. Chang, *Tetrahedron Lett.*, 1990, **31**, 4765.
56. G. A. Molander, B. Czakó and M. Rheam, *J. Org. Chem.*, 2007, **72**, 1755.
57. G. A. Molander and J. A. McKie, *J. Org. Chem.*, 1994, **59**, 3186.
58. G. A. Molander, K. M. George and L. G. Monovich, *J. Org. Chem.*, 2003, **68**, 9533.
59. K. Weinges, S. B. Schmidbauer and H. Schick, *Chem. Ber.*, 1994, **127**, 1305.
60. D. Johnston, C. M. McCusker and D. J. Procter, *Tetrahedron Lett.*, 1999, **40**, 4913.
61. D. Johnston, C. F. McCusker, K. Muir and D. J. Procter, *J. Chem. Soc., Perkin Trans. 1*, 2000, 681.
62. E. J. Enholm, H. Satici and A. Trivellas, *J. Org. Chem.*, 1989, **54**, 5841.
63. K. C. Nicolaou, A. Li and D. J. Edmonds, *Angew. Chem. Int. Ed.*, 2006, **45**, 7086.
64. K. Suzuki and T. Nakata, *Org. Lett.*, 2002, **4**, 3943.
65. G. A. Molander, J. C. McWilliams and B. C. Noll, *J. Am. Chem. Soc.*, 1997, **119**, 1265.
66. H.-G. Schmalz, S. Siegel and J. W. Bats, *Angew. Chem. Int. Ed. Engl.*, 1995, **34**, 2383.
67. H.-G. Schmalz, S. Siegel and A. Schwarz, *Tetrahedron Lett.*, 1996, **37**, 2947.
68. F. Aulenta, M. Berndt, I. Brüdgam, H. Hartl, S. Sörgel and H.-U. Reissig, *Chem. Eur. J.*, 2007, **13**, 6047.
69. S. Gross and H.-U. Reissig, *Org. Lett.*, 2003, **5**, 4305.
70. D. Riber and T. Skrydstrup, *Org. Lett.*, 2003, **5**, 229.
71. S. A. Johannesen, S. Albu, R. G. Hazell and T. Skrydstrup, *Chem. Commun.*, 2004, 1962.
72. G. Masson, P. Cividino, S. Py and Y. Vallée, *Angew. Chem. Int. Ed.*, 2003, **42**, 2265.
73. P. Cividino, S. Py, P. Delair and A. E. Greene, *J. Org. Chem.*, 2007, **72**, 485.
74. S. Desvergnes, S. Py and Y. Vallée, *J. Org. Chem.*, 2005, **70**, 1459.
75. H. M. Peltier, J. P. McMahon, A. W. Patterson and J. A. Ellman, *J. Am. Chem. Soc.*, 2006, **128**, 16018.
76. E. Hasegawa and D. P. Curran, *Tetrahedron Lett.*, 1993, **34**, 1717.
77. S. Fukuzawa and T. Tsuchimoto, *Synlett*, 1993, 803.

78. P. de Pouilly, A. Chénedé, J.-M. Mallet and P. Sinaÿ, *Tetrahedron Lett.*, 1992, **33**, 8065.
79. T. Skrydstrup, D. Mazéas, M. Elmouchir, G. Doisneau, C. Riche, A. Chiaroni and J.-M. Beau, *Chem. Eur. J.*, 1997, **3**, 1342.
80. D. Mazéas, T. Skrydstrup, O. Doumeix and J.-M. Beau, *Angew. Chem. Int. Ed. Engl.*, 1994, **33**, 1383.
81. G. A. Molander and C. R. Harris, *J. Org. Chem.*, 1997, **62**, 7418.
82. S. M. Bennett, R. K. Biboutou, Z. H. Zhou and R. Pion, *Tetrahedron*, 1998, **54**, 4761.
83. B. S. F. Salari, R. K. Biboutou and S. M. Bennett, *Tetrahedron*, 2000, **56**, 6385.
84. T. J. K. Findley, D. Sucunza, L. C. Miller, D. T. Davies and D. J. Procter, *Chem. Eur. J.*, 2008, **14**, 6862.
85. J. M. Aurrecoechea and A. Fernández-Acebes, *Tetrahedron Lett.*, 1993, **34**, 549.
86. C. W. Holzapfel and D. B. G. Williams, *Tetrahedron*, 1995, **51**, 8555.
87. Y.-Y. Jaing, Q. Li, W. Lu and J.-C. Cai, *Tetrahedron Lett.*, 2003, **44**, 2073.
88. J. Inanaga, J. Katsuki, O. Ujikawa and M. Yamaguchi, *Tetrahedron Lett.*, 1991, **32**, 4921.
89. G. A. Molander and C. R. Harris, *J. Am. Chem. Soc.*, 1996, **118**, 4059.
90. Y. Aoyagi, T. Inariyama, Y. Arai, S. Tsuchida, Y. Matuda, H. Kobayashi and A. Ohta, *Tetrahedron*, 1994, **50**, 13575.
91. Y. Aoyagi, T. Manabe, A. Ohta, T. Kurihara, G.-L. Pang and T. Yuhara, *Tetrahedron*, 1996, **52**, 869.
92. J. Inanaga, O. Ujikawa and M. Yamaguchi, *Tetrahedron Lett.*, 1991, **32**, 1737.
93. G. A. Molander and L. S. Harring, *J. Org. Chem.*, 1990, **55**, 6171.
94. C. F. Sturino and A. G. Fallis, *J. Org. Chem.*, 1994, **59**, 6514.
95. J. J. C. Grové and C. W. Holzapfel, *Tetrahedron Lett.*, 1997, **38**, 7429.
96. J. Marco-Contelles, P. Gallego, M. Rodríguez-Fernández, N. Khiar, C. Destabel, M. Bernabé, A. Martínez-Grau and J. L. Chiara, *J. Org. Chem.*, 1997, **62**, 7397.
97. A. Krief and A.-M. Laval, *Chem. Rev.*, 1999, **99**, 745.
98. P. Girard, J.-L. Namy and H. B. Kagan, *J. Am. Chem. Soc.*, 1980, **102**, 2693.
99. K. Otsubo, K. Kawamura, J. Inanaga and M. Yamaguchi, *Chem. Lett.*, 1987, 1487.
100. F. Machrouhi, B. Hamann, J.-L. Namy and H. B. Kagan, *Synlett*, 1996, 633.
101. G. A. Molander and C. Alonso-Alija, *J. Org. Chem.*, 1998, **63**, 4366.
102. G. A. Molander and F. Machrouhi, *J. Org. Chem.*, 1999, **64**, 4119.
103. G. A. Molander and J. B. Etter, *J. Org. Chem.*, 1986, **51**, 1778.
104. R. S. Miller, J. M. Sealy, M. Shabangi, M. L. Kuhlman, J. R. Fuchs and R. A. Flowers II, *J. Am. Chem. Soc.*, 2000, **122**, 7718.
105. G. A. Molander and C. R. Harris, *Tetrahedron*, 1998, **54**, 3321.
106. H.-Y. Kang and S.-E. Song, *Tetrahedron Lett.*, 2000, **41**, 937.

107. B. Hamann-Gaudinet, J.-L. Namy and H. B. Kagan, *Tetrahedron Lett.*, 1997, **38**, 6585.
108. D. P. Curran, T. L. Fevig, C. P. Jasperse and M. J. Totleben, *Synlett*, 1992, 943.
109. M. Kunishima, K. Hioki, K. Kono, T. Sakuma and S. Tani, *Chem. Pharm. Bull.*, 1994, **42**, 2190.
110. M. J. Totleben, D. P. Curran and P. Wipf, *J. Org. Chem.*, 1992, **57**, 1740.
111. M.-I. Lannou, F. Hélion and J.-L. Namy, *Tetrahedron*, 2003, **59**, 10551.
112. D. P. Curran and M. J. Totleben, *J. Am. Chem. Soc.*, 1992, **114**, 6050.
113. G. A. Molander and J. A. McKie, *J. Org. Chem.*, 1991, **56**, 4112.
114. D. P. Curran, X. Gu, W. Zhang and P. Dowd, *Tetrahedron*, 1997, **53**, 9023.
115. M. Murakami, M. Hayashi and Y. Ito, *J. Org. Chem.*, 1992, **57**, 793.
116. M. Kunishima, S. Tanaka, K. Kono, K. Hioki and S. Tani, *Tetrahedron Lett.*, 1995, **36**, 3707.
117. M. Kunishima, D. Nakata, S. Tanaka, K. Hioki and S. Tani, *Tetrahedron*, 2000, **56**, 9927.
118. M. Kunishima, K. Yoshimura, D. Nakata, K. Hioki and S. Tani, *Chem. Pharm. Bull.*, 1999, **47**, 1196.
119. S. W. Youn, H. S. Park and Y. H. Kim, *Chem. Commun.*, 2000, 2005.
120. T. Imamoto, T. Hatajima, N. Takiyama, T. Takeyama, Y. Kamiya and T. Yoshizawa, *J. Chem. Soc., Perkin Trans. 1*, 1991, 3127.
121. J. D. White and T. C. Somers, *J. Am. Chem. Soc.*, 1987, **109**, 4424.
122. T. Tabuchi, J. Inanaga and M. Yamaguchi, *Tetrahedron Lett.*, 1986, **27**, 3891.
123. S. Matsubara, M. Yoshioka and K. Utimoto, *Angew. Chem. Int. Ed. Engl.*, 1997, **36**, 617.
124. J. M. Concellón, P. L. Bernad and J. A. Pérez-Andrés, *Tetrahedron Lett.*, 1998, **39**, 1409.
125. M. Yoshida, D. Suzuki and M. Iyoda, *J. Chem. Soc., Perkin Trans. 1*, 1997, 643.
126. X. Zheng, C.-G. Feng, J.-L. Ye and P.-Q. Huang, *Org. Lett.*, 2005, **7**, 553.
127. D. R. Williams, M. A. Berliner, B. W. Stroup, P. P. Nag and M. P. Clark, *Org. Lett.*, 2005, **7**, 4099.
128. S.-i. Fukuzawa, H. Furuya and T. Tsuchimoto, *Tetrahedron*, 1996, **52**, 1953.
129. F. Matsuda, T. Sakai, N. Okada and M. Miyashita, *Tetrahedron Lett.*, 1998, **39**, 863.
130. M. Kito, T. Sakai, H. Shirahama, M. Miyashita and F. Matsuda, *Synlett*, 1997, 219.
131. H. Tamiya, K. Goto and F. Matsuda, *Org. Lett.*, 2004, **6**, 545.
132. G. A. Molander, J. B. Etter and P. W. Zinke, *J. Am. Chem. Soc.*, 1987, **109**, 453.
133. G. Lannoye, K. Sambasivarao, S. Wehrli and J. M. Cook, *J. Org. Chem.*, 1998, **53**, 2327.

134. G. A. Molander, Y. L. Huérou and G. A. Brown, *J. Org. Chem.*, 2001, **66**, 4511.
135. G. A. Molander and J. A. McKie, *J. Org. Chem.*, 1993, **58**, 7216.
136. G. A. Molander and S. R. Shakya, *J. Org. Chem.*, 1994, **59**, 3445.
137. A. Fadel, *Tetrahedron: Asymmetry*, 1994, **5**, 531.
138. D.-C. Ha, C.-S. Yun and E. Yu, *Tetrahedron Lett.*, 1996, **37**, 2577.
139. I. M. Rudkin, L. C. Miller and D. J. Procter, *Organomet. Chem.*, 2008, **34**, 19.
140. S.-i. Fukuzawa, H. Matsuzawa and S.-i. Yoshimitsu, *J. Org. Chem.*, 2000, **65**, 1702.
141. J. M. Concellón, C. Concellón and C. Mejica, *J. Org. Chem.*, 2005, **70**, 6111.
142. J. M. Concellón, C. Concellón and P. Diaz, *Eur. J. Org. Chem.*, 2006, 2197.
143. J. M. Concellón and C. Concellón, *J. Org. Chem.*, 2006, **71**, 4428.
144. K. Utimoto, T. Matsui, T. Takai and S. Matsubara, *Chem. Lett.*, 1995, 197.
145. A. Otaka, J. Watanabe, A. Yukimasa, Y. Sasaki, H. Watanabe, T. Kinoshita, S. Oishi, H. Tamamura and N. Fujii, *J. Org. Chem.*, 2004, **69**, 1634.
146. S. N. Baytas, Q. Wang, N. A. Karst, J. S. Dordick and R. L. Linhardt, *J. Org. Chem.*, 2004, **69**, 6900.
147. Y. Aoyagi, M. Yoshimura, M. Tsuda, T. Tsuchibuchi, S. Kawamata, H. Tateno, K. Asano, H. Nakamura, M. Obokata, A. Ohta and Y. Kodama, *J. Chem. Soc., Perkin Trans. 1*, 1995, 689.
148. Y. Aoyagi, R. Asakura, N. Kondoh, R. Yamamoto, T. Kuromatsu, A. Shimura and A. Ohta, *Synthesis*, 1996, 970.
149. T. Tabuchi, K. Kawamura, J. Inanaga and M. Yamaguchi, *Tetrahedron Lett.*, 1986, **27**, 3889.
150. J. Inanaga, Y. Yokoyama, Y. Handa and M. Yamaguchi, *Tetrahedron Lett.*, 1991, **32**, 6371.
151. G. A. Molander, J. B. Etter, L. S. Harring and P.-J. Thorel, *J. Am. Chem. Soc.*, 1991, **113**, 8036.
152. T. Mukaiyama, I. Shiina, H. Iwadare, M. Saitoh, T. Nishimura, N. Ohkawa, H. Sakoh, K. Nishimura, Y.-I. Tani, E. Hasegawa, K. Yamada and K. Saitoh, *Chem. Eur. J.*, 1999, **1**, 121.
153. E. J. Enholm, S. Jiag and K. Abboud, *J. Org. Chem.*, 1993, **58**, 4061.
154. F. McKerlie, I. M. Rudkin, G. Wynne and D. J. Procter, *Org. Biomol. Chem.*, 2005, **3**, 2805.
155. S. Ichikawa, S. Shuto and A. Matsuda, *J. Am. Chem. Soc.*, 1999, **121**, 10270.
156. T. Kodama, S. Shuto, S. Ichikawa and A. Matsuda, *J. Org. Chem.*, 2002, **67**, 7706.
157. M. Ricci, P. Blakskjaer and T. Skrydstrup, *J. Am. Chem. Soc.*, 2000, **122**, 12413.
158. T. Mukaiyama, H. Arai and I. Shiina, *Chem. Lett.*, 2000, 580.

159. T. Mukaiyama, K. Pudhom, K. Yamane and H. Arai, *Bull. Chem. Soc. Jpn.*, 2003, **76**, 413.
160. Y. Ogawa, K. Kuroda and T. Mukaiyama, *Chem. Lett.*, 2005, **34**, 372.
161. Y. Ogawa, K. Kuroda and T. Mukaiyama, *Chem. Lett.*, 2005, **34**, 698.
162. M. F. Jacobsen, M. Turks, R. Hazell and T. Skrydstrup, *J. Org. Chem.*, 2002, **67**, 2411.
163. A. Cabrera, R. Le Lagadec, P. Sharma, J. L. Arias, R. A. Toscano, L. Velasco, R. Gaviño, C. Alvarez and M. Salmón, *J. Chem. Soc., Perkin Trans. 1*, 1998, 3609.
164. S.-J. Jong and J.-M. Fang, *Org. Lett.*, 2000, **2**, 1947.
165. G. Guazzelli, L. A. Duffy and D. J. Procter, *Org. Lett.*, 2008, **19**, 4291.
166. E. J. Enholm and A. Trivellas, *Tetrahedron Lett.*, 1994, **35**, 1627.
167. G. A. Molander, G. A. Brown and I. Storch de Gracia, *J. Org. Chem.*, 2002, **67**, 3459.

CHAPTER 6

Sequential Carbon–Carbon Bond Formation Using SmI_2

6.1 An Introduction to Sequential Reactions Mediated by SmI_2

Sequential reactions, in which a number of transformations convert simple starting materials to complex products, using a single reagent, in a single synthetic operation, are the holy grail of the synthetic chemist. Individual reactions in the sequence must be highly selective if the desired products are to be obtained efficiently from the reaction cascade. Unfortunately, few reagents are able to meet this challenge. Of the many reducing agents available to the synthetic chemist, SmI_2 is the one reagent able to orchestrate powerful sequential processes, often with exquisite control of structure and stereochemistry. Molander has pioneered the use of SmI_2 in sequential reactions and has reviewed the subject.[1,2] In addition, Skrydstrup has discussed the emerging area in a highlight.[3] This chapter aims to provide illustrative examples of sequential reactions mediated by SmI_2 and to relate the component parts of each sequence to discussions of mechanism and reactivity in preceding chapters. The chapter focuses on examples that result in *sequential carbon–carbon bond formation* and a dramatic increase in structural complexity.

6.2 Sequences Initiated by Radical Cyclisations

6.2.1 Radical–Radical Sequences

Curran's elegant approach to the natural products (±)-hypnophilin and (±)-coriolin is among the best examples of a SmI_2-mediated sequence and consists of a 5-*exo-trig* cyclisation followed by a stereoselective 5-*exo-dig* cyclisation in a radical–radical cascade (Chapter 7, Section 7.4).[4] More recently, Kilburn employed a SmI_2-mediated radical–radical sequence involving a methylenecyclopropyl ketone in the preparation of paeonilactone B.[5,6] In

Organic Synthesis Using Samarium Diiodide: A Practical Guide
By David J. Procter, Robert A. Flowers, II and Troels Skrydstrup

Published by the Royal Society of Chemistry, www.rsc.org

Kilburn's sequential approach, the six- and five-membered rings of the target are generated stereoselectively in one synthetic operation. Treatment of methylenecyclopropyl ketone **1** with SmI_2, in the presence of HMPA and *t*-BuOH, gave the bicyclic ether **2** as a 10:1 mixture of diastereoisomers (Scheme 6.1).

Representative procedure. To freshly prepared SmI_2 (2.2 equiv, 0.15 M in THF) HMPA (21.4 equiv) was added and the solution was cooled to 0 °C. Methylenecyclopropyl ketone **1** (1 equiv) and *t*-BuOH (2 equiv) as a solution in THF were added to the reaction mixture dropwise over 90 min. The reaction mixture was allowed to warm to room temperature before quenching with 5% aqueous citric acid and extracting into petroleum ether–EtOAc (1:1). The organic layers were washed with brine, dried ($MgSO_4$) and concentrated *in vacuo*. Crude **2** was purified by column chromatography (petroleum ether–EtOAc).

The reaction mechanism involved the cyclisation of ketyl radical anion **3** on to the methylenecyclopropane moiety in a 5-*exo-trig* manner. Ring opening of cyclopropane intermediate **4** gave rise to the cyclohexyl radical **5**, which then cyclised in a 5-*exo-dig* fashion to form the second ring (Scheme 6.2).[5,6]

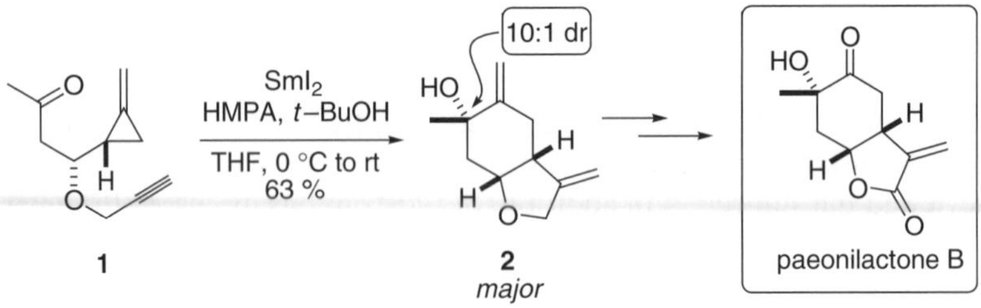

Scheme 6.1

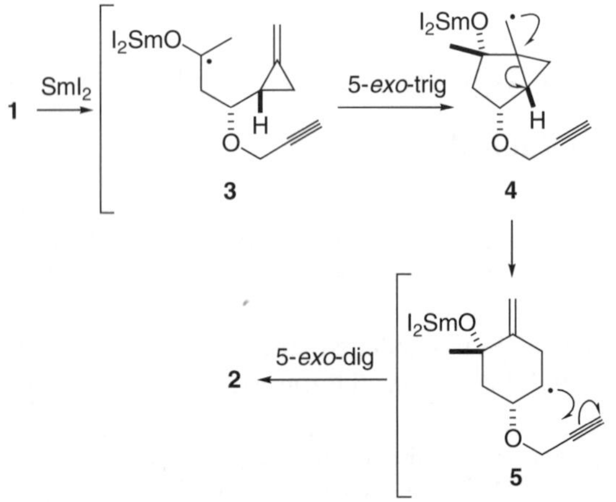

Scheme 6.2

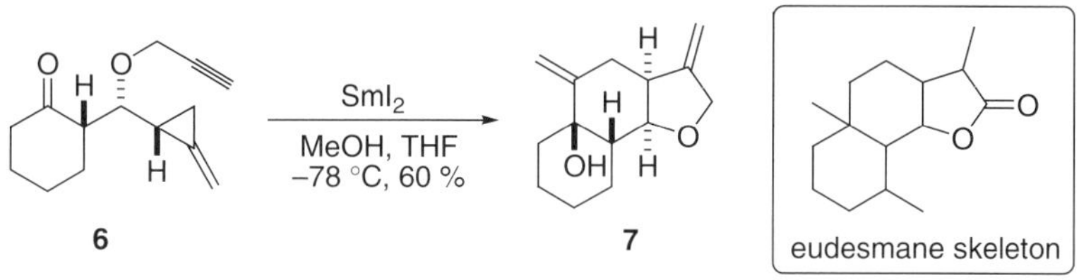

Scheme 6.3

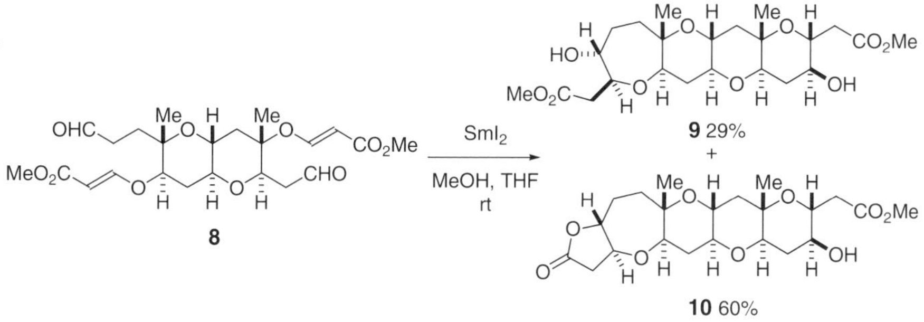

Scheme 6.4

A related radical–radical sequence was used by Kilburn to access the tricyclic ether skeleton of the eudesmanes.[7] It was found that SmI_2 in THF–MeOH (4:1) gave the most efficient conversion of methylenecyclopropyl ketone **6** to tricycle **7** (Scheme 6.3).

Finally, Nakata recently reported the application of a radical–radical sequence in a two-directional approach to the B–E ring domain of maitotoxin.[8] Treatment of dialdehyde **8** with SmI_2 and MeOH gave lactone **9** and hydroxy ester **10** in excellent overall yield (Scheme 6.4).[8]

Representative procedure. To dialdehyde **8** (1 equiv) and MeOH (8 equiv) in THF was added SmI_2 (12 equiv, 0.1 M in THF) at room temperature under argon. The reaction mixture was stirred for 1 h before the addition of aqueous saturated $Na_2S_2O_3$ at 0 °C. The reaction product was extracted into EtOAc and the organic layers were washed with brine, dried ($MgSO_4$) and concentrated *in vacuo*. Crude **9** and **10** were purified by column chromatography (hexane–EtOAc).

6.2.2 Radical–Anionic Sequences

When planning sequences that require a radical cyclisation, care must be taken as radical intermediates are rapidly reduced by SmI_2 to give organosamariums. Curran has provided rate data for the reduction of primary, secondary and tertiary radicals to anions (Chapter 3, Section 3.2).[9] The desired radical cyclisation must be faster than reduction of the radical to the corresponding anion [$>(5–7)\times 10^6\,M^{-1}\,s^{-1}$ for a primary radical, depending on HMPA concentration].[9] Provided that this is the case, radical cyclisation–anionic sequences can be planned and realised. For example, treatment of iodide **11** with SmI_2 in the presence of 2-methylcyclohexanone gave **12**, the product of an efficient 5-*exo-trig* radical cyclisation–intermolecular Barbier addition (Scheme 6.5).[10] The use of activated alkenes facilitates 5-*exo-trig* radical cyclisation; for example, cyclisation of iodide **13** gave rise to a Sm(III) enolate that underwent an intermolecular Claisen reaction to give **14** (Scheme 6.5).[1,2] Finally, radical–anionic sequences involving more challenging 6-*exo-trig* cyclisations are sometimes possible. For example, treatment of aryl iodide **15** with SmI_2 in the presence of cyclohexanone gave **16** after a 6-*exo-trig* radical cyclisation–intermolecular Barbier sequence (Scheme 6.5).[10]

Curran described an efficient, approach to the BCD ring system of the penitrem indole alkaloids that exploits a SmI_2-mediated radical–anionic sequence.[11] Treatment of iodide **17** with SmI_2 and HMPA gave **18**, the product of radical cyclisation–intermolecular Barbier addition, in satisfactory yield (Scheme 6.6).

radical cyclisation/Barbier addition sequence

11 + 2-methylcyclohexanone → SmI_2, THF, HMPA, rt, 74% → **12**

radical cyclisation/Claisen addition sequence

13 + PhC(O)OMe → SmI_2, cat. NiI_2, THF, 0 °C, 51% → **14**

radical cyclisation/Barbier addition sequence

15 + cyclohexanone → SmI_2, THF, HMPA, rt, 57% → **16**

Scheme 6.5

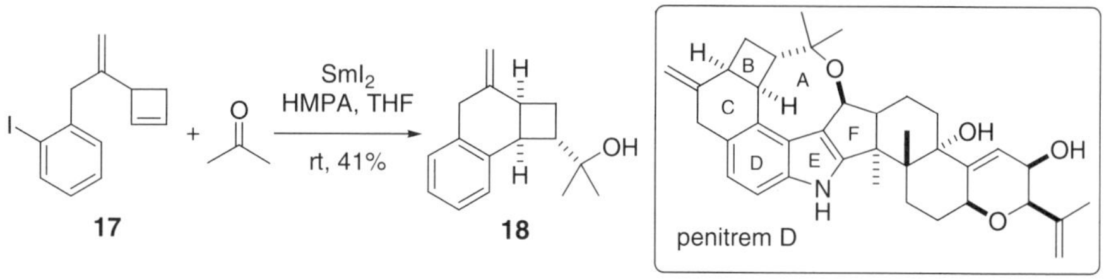

Scheme 6.6

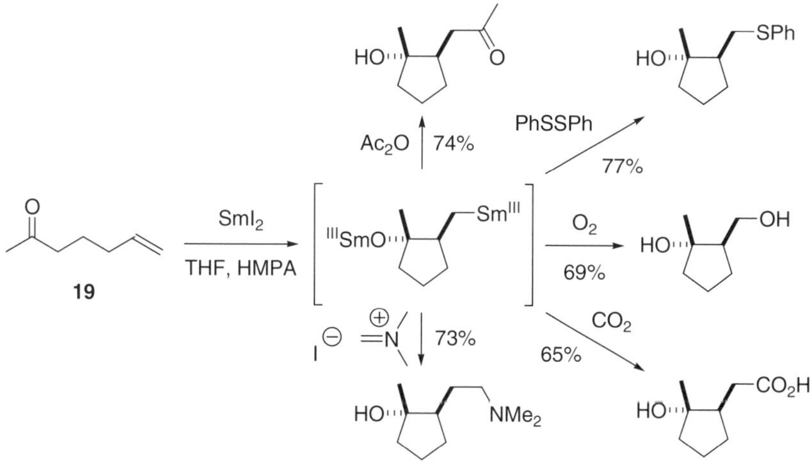

Scheme 6.7

Radical cyclisation–anionic sequences, in which the radical cyclisation involves the addition of a ketyl radical anion to an alkene, are also possible. For example, carbonyl–alkene cyclisation of unsaturated ketone **19** and quenching of the resultant organosamarium with electrophiles allow access to an impressive range of cyclopentanol products (Scheme 6.7).[12]

Chelation-controlled ketyl radical cyclisations can also be employed in radical–anionic sequences. For example, treatment of ketolactone **20** with SmI_2 in the presence of acetone gives adduct **21** with good diastereocontrol (Scheme 6.8).[13]

Representative procedure. To SmI_2 (2.1 equiv, 0.1 M in THF) at −30 °C were added ketolactone **20** (1 equiv) and dry acetone (1.5 equiv) as a solution in THF. The reaction mixture was allowed to warm to room temperature over 3 h before quenching with saturated aqueous $NaHCO_3$ and extracting into Et_2O. The organic layers were washed with brine, dried (Na_2SO_4) and concentrated *in vacuo*. Crude **21** was purified by column chromatography (hexane–EtOAc).

Activated alkenes can also be employed in radical–anionic sequences involving ketyl radical cyclisations; in these cases, the anionic intermediates are

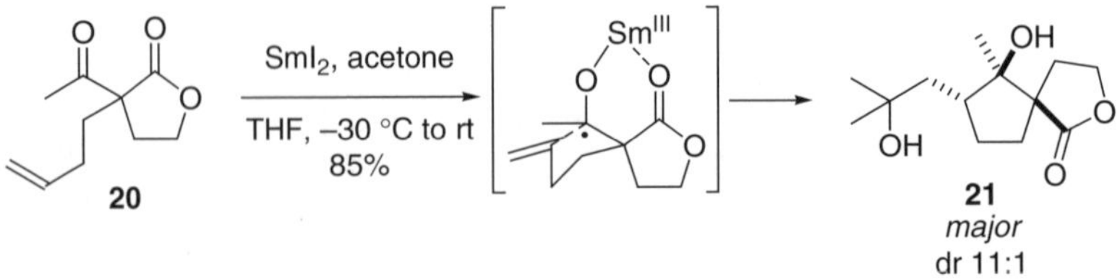

Scheme 6.8

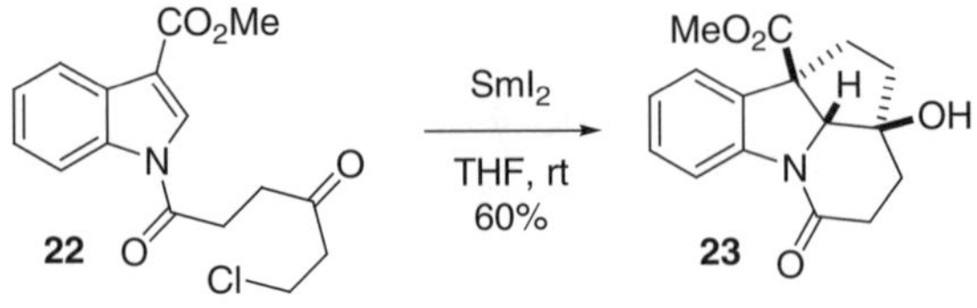

Scheme 6.9

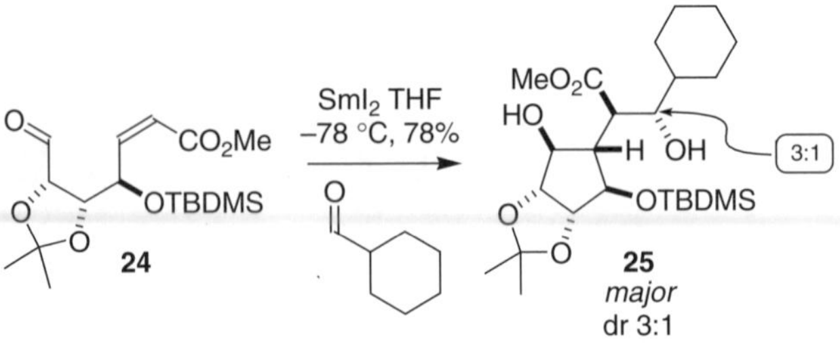

Scheme 6.10

often Sm(III) enolates. For example, Reissig has reported radical–anionic sequences that involve ketyl radical cyclisations on to indoles.[14] Dearomatising radical cyclisation results in the formation of a Sm(III) enolate that can be alkylated *in situ* with allyl halides (see Chapter 5, Section 5.2). In one example, Reissig reported the interception of the Sm(III) enolate intermediate with an internal alkyl halide: treatment of indole **22** with SmI_2 gave tetracyclic product **23** in good yield and as a single diastereoisomer (Scheme 6.9).[14]

Enholm reported radical–anionic sequences involving ketyl radical cyclisations that culminate in intermolecular aldol reactions: treatment of aldehyde **24** and cyclohexane carboxaldehyde with SmI_2 triggers a radical cyclisation–intermolecular aldol sequence to give **25** in good yield (Scheme 6.10).[15]

Representative procedure. To SmI_2 (0.1 M in THF, 3 equiv) at –78 °C were added aldehyde **24** (1 equiv) and cyclohexanecarboxaldehyde (2 equiv) slowly using a syringe pump. The product was isolated by column chromatography.

SmI_2

t-BuOH, THF

0 °C

only diastereoisomer

26 R = –C(Me)=CH_2
27 R = 3–butenyl

28 R = –C(Me)=CH_2, 90%
29 R = 3–butenyl, 72%

anti–radical cyclisation

30

chelation–controlled aldol cyclisation

Scheme 6.11

Procter has recently shown that radical cyclisation–*intramolecular* aldol sequences can also be carried out.[16] For example, treatment of dialdehydes **26** and **27** with SmI_2 results in a sequential cyclisation to give **28** and **29**, respectively, as single diastereoisomers in high yield (Scheme 6.11). The diastereoselectivity of the sequence can be ascribed to the chelation of Sm(III) to intermediates, most notably in controlling the double-bond stereochemistry during the formation of Sm(III) enolate **30**.[16]

Perhaps the most surprising feature of the sequence is that selective reduction of one aldehyde is possible while the other survives to act as an electrophile in the later aldol cyclisation. The necessity for pre-coordination between the aldehyde and the ester carbonyl group of the alkene acceptor prior to electron transfer to the aldehyde could be responsible for this selectivity.[16]

6.3 Sequences Initiated by Anionic Reactions

6.3.1 Anionic–Radical and Anionic–Anionic Sequences

Molander recognised the potential of the SmI_2-mediated Barbier addition to esters for the initiation of sequential processes (Chapter 5, Section 5.4). Two types of cascade have been developed that involve nucleophilic acyl substitution: the first type involves double intramolecular Barbier addition to an ester group (anionic–anionic sequences),[17] and the second type consists of a Barbier addition to an ester followed by a carbonyl–alkene/alkyne cyclisation of the resultant ketone (anionic–radical sequences) (Scheme 6.12).[18,19]

The sequential nucleophilic substitution–Barbier additions are controlled by the different rates of reduction of the alkyl halides or the tether length, if the

– SmI_2–mediated nucleophilic acyl substitution, Barbier addition
(an anionic/anionic sequence)

SmI_2 (4 equiv)
THF, HMPA
0 °C to rt
67%

– SmI_2–mediated nucleophilic acyl substitution, carbonyl–alkene/alkyne cyclisation
(an anionic/radical sequence)

SmI_2 (4 equiv)
THF, HMPA
0 °C to rt
81%

Scheme 6.12

same halides are present in the substrate.[17] Molander showed that anionic–radical sequences are effective for the synthesis of a range of bicyclic oxygen heterocycles.[19] For example, treatment of iodide **31** with SmI_2–HMPA triggers formation of cyclobutanone intermediate **32**, which then undergoes radical cyclisation to give tetrahydrofuran **33** in good yield (Scheme 6.13).[19]

Representative procedure.[19] To a solution of SmI_2 (5.5 equiv) in THF was added HMPA (25 equiv) and the solution was cooled to 0 °C. The iodoester (1 equiv) in dry THF was the added dropwise using a cannula over 2 h. The reaction was monitored by TLC and, once complete, the mixture was allowed to warm to room

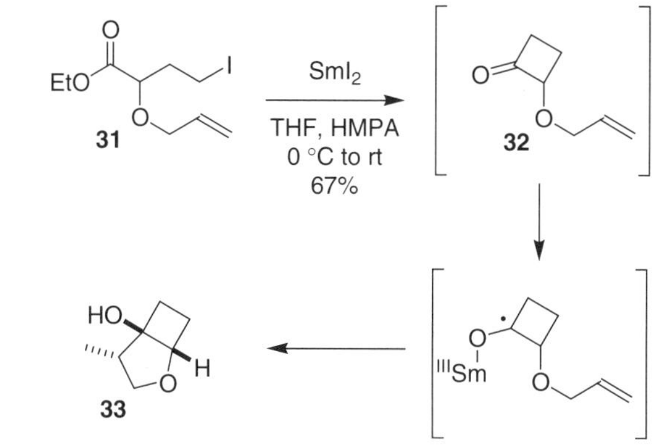

Scheme 6.13

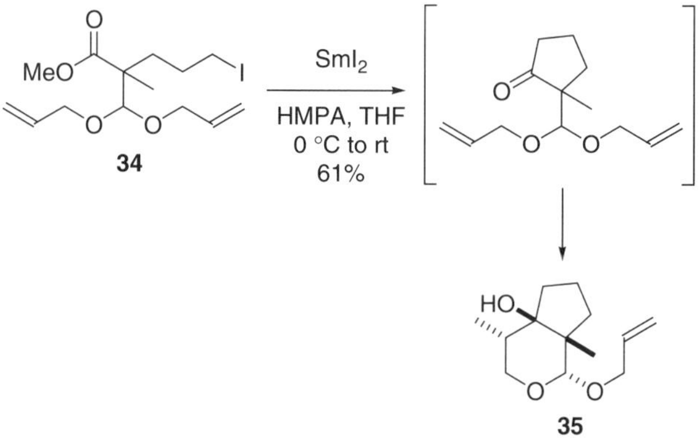

Scheme 6.14

temperature before quenching the reaction with saturated $NaHCO_3$. The aqueous layer was extracted and the organic extracts were dried and concentrated. The crude product was purified by column chromatography (EtOAc–hexanes eluent).

Acetals such as **34** have also been used as substrates in anionic–radical cascades: treatment of **34** with SmI_2–HMPA gives bicyclic acetal **35** with excellent stereocontrol (Scheme 6.14).[19]

Molander reported a variant of this sequential process that results in alkenyl transfer.[20] For example, treatment of bromide **36** with SmI_2–HMPA resulted in the formation of cyclopentanone **37**. Subsequent intramolecular ketyl radical addition to the enol ether and collapse of the resultant organosamarium intermediate **38** gave cyclopentanol **39** in good overall yield (Scheme 6.15).[20]

These anionic–radical sequences can even be extended further and Molander has shown that anionic–radical–anionic sequences are possible.[18] Treatment of

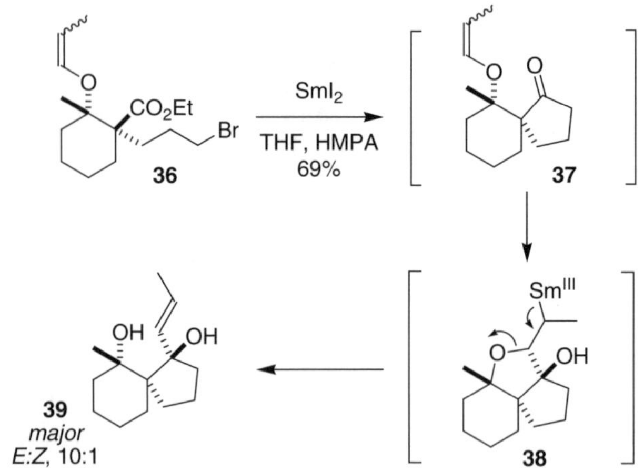

Scheme 6.15

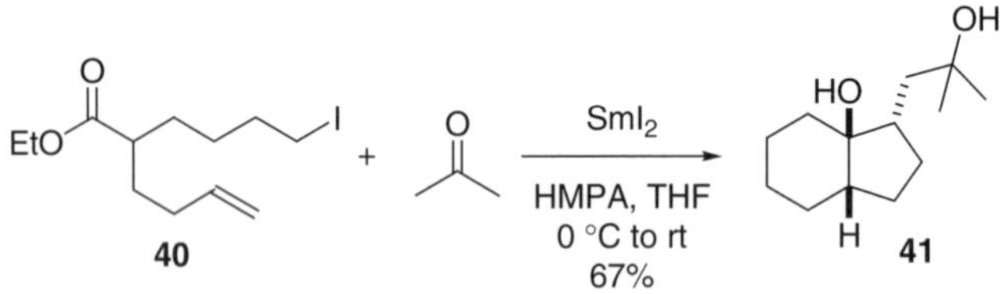

Scheme 6.16

40 with SmI_2 and HMPA in the presence of acetone gave bicyclic alcohol **41** in good yield (Scheme 6.16).

Anionic–anionic cascades are also effective for the rapid assembly of carbon skeletons. For example, double Barbier cyclisations were used by Cook for the synthesis of tetracyclic polyquinenes.[21] Molander has shown that such sequences can be used to access a range of fused polycyclic systems (Scheme 6.17).[13]

Molander has also studied the SmI_2-mediated double Barbier additions of alkyl dihalides to ketoesters.[22,23] These impressive anionic–anionic, intermolecular–intramolecular sequences require the use of NiI_2 as an additive and irradiation with visible light and allow access to a range of bicyclic and tricyclic systems. The reactions proceed by reduction of the more reactive alkyl halide, intermolecular Barbier addition to the ketone, lactonisation and a second Barbier addition to the lactone carbonyl (Scheme 6.18).[22]

Representative procedure.[22] To SmI_2 (7 equiv) in THF was added NiI_2 (1.2 equiv) and the solution was cooled to 0 °C and stirred for 5 min. The ester-ketone (1 equiv) was then added, followed immediately by dropwise addition of the dihalide

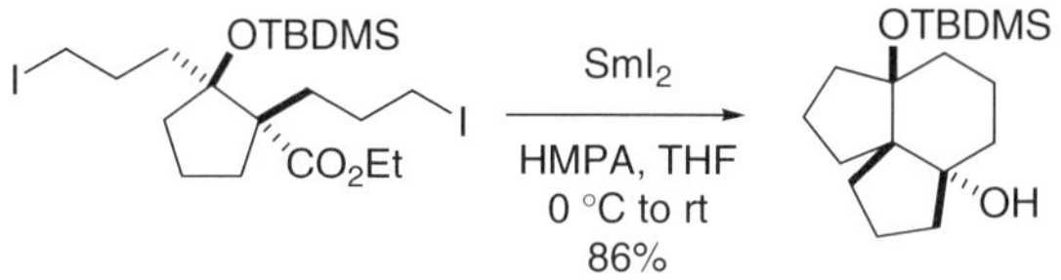

Scheme 6.17

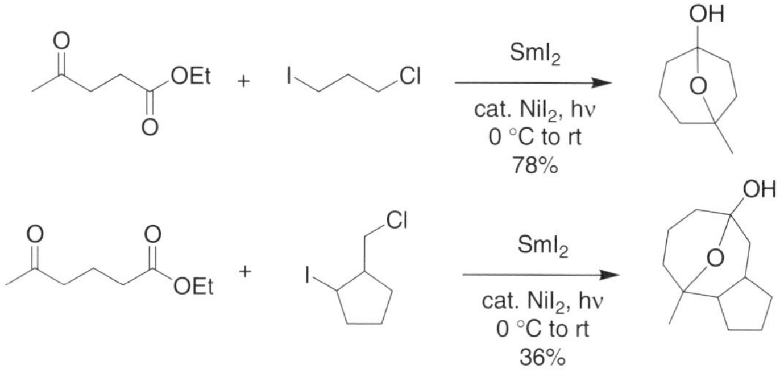

Scheme 6.18

(1.1 equiv) over 30 min. Once the addition was complete, the reaction was irradiated with visible light (250 W krypton lamp) for 3 h while keeping the temperature below 25 °C. The resultant mixture was quenched with Rochelle's salt and 10% aqueous K_2CO_3 followed by extraction with Et_2O. The combined organic layers were washed with brine and dried ($MgSO_4$) and the solvent was evaporated to yield the crude product, which was purified by column chromatography (EtOAc–hexanes eluent) and distillation.

References

1. G. A. Molander and C. R. Harris, *Chem. Rev.*, 1996, **96**, 307.
2. G. A. Molander and C. R. Harris, *Tetrahedron*, 1998, **54**, 3321.
3. T. Skrydstrup, *Angew. Chem. Int. Ed. Engl.*, 1997, **36**, 345.
4. T. L. Fevig, R. L. Elliot and D. P. Curran, *J. Am. Chem. Soc.*, 1988, **110**, 5064.
5. R. J. Boffey, M. Santagostino, W. G. Whittingham and J. D. Kilburn, *J. Chem. Soc., Chem. Commun.*, 1998, 1875.
6. R. J. Boffey, W. G. Whittingham and J. D. Kilburn, *J. Chem. Soc., Perkin Trans. 1*, 2001, 487.
7. F. C. Watson and J. D. Kilburn, *Tetrahedron Lett.*, 2000, **41**, 10341.
8. M. Satoh, H. Koshino and T. Nakata, *Org. Lett.*, 2008, **10**, 1683.
9. E. Hasegawa and D. P. Curran, *Tetrahedron Lett.*, 1993, **34**, 1717.
10. G. A. Molander and L. S. Harring, *J. Org. Chem.*, 1990, **55**, 6171.
11. A. Rivkin, T. Nagashima and D. P. Curran, *Org. Lett.*, 2003, **5**, 419.

12. G. A. Molander and J. A. McKie, *J. Org. Chem.*, 1992, **57**, 3132.
13. G. A. Molander and C. Kenny, *J. Org. Chem.*, 1991, **56**, 1439.
14. S. Gross and H.-U. Reissig, *Org. Lett.*, 2003, **4**, 4305.
15. E. J. Enholm and A. Trivellas, *Tetrahedron Lett.*, 1994, **35**, 1627.
16. M. D. Helm, D. Sucunza, M. Da Silva, M. Helliwell and D. J. Procter, *Tetrahedron Lett.*, 2009, **50**, 3224.
17. G. A. Molander and C. R. Harris, *J. Am. Chem. Soc.*, 1995, **117**, 3705.
18. G. A. Molander and C. R. Harris, *J. Am. Chem. Soc.*, 1996, **118**, 4059.
19. G. A. Molander and C. R. Harris, *J. Org. Chem.*, 1997, **62**, 2944.
20. G. A. Molander and C. R. Harris, *J. Org. Chem.*, 1998, **63**, 4374.
21. G. Lannoye, K. Sambasivarao, S. Wehrli and J. M. Cook, *J. Org. Chem.*, 1998, **53**, 2327.
22. G. A. Molander and C. Alonso-Alija, *J. Org. Chem.*, 1998, **63**, 4366.
23. G. A. Molander and F. Machrouhi, *J. Org. Chem.*, 1999, **64**, 4119.

CHAPTER 7

Emerging Areas

7.1 Other Sm(II) Reagents

Although the focus of this book is on organic reactions mediated by SmI_2, the reagent is also very useful for the synthesis of a wide range of Sm(II)-based reductants (Scheme 7.1). This section will briefly describe the synthesis and utility of these reagents.

7.1.1 Sm(II) Chloride and Sm(II) Bromide

Although SmI_2 is by far the most commonly used reductant among Sm(II) halides, samarium(II) chloride ($SmCl_2$) and samarium(II) bromide ($SmBr_2$) have been used in a number of important synthetic applications. The main shortcoming of these reagents is their low solubility in most organic solvents. Nonetheless, there are a number of instances where their use provides advantages over SmI_2.

The synthesis of $SmBr_2$ was first reported in 1934 by Selwood through the reduction of $SmBr_3$ with H_2 at 740 °C.[1] More recently, Kagan developed a preparation of $SmBr_2$ through the conversion of Sm_2O_3 to a $SmBr_3$ hydrate followed by drying and reduction with a lithium dispersion in THF.[2] Since Kagan's report, a number of other approaches to $SmBr_2$ have been developed. Flowers showed that addition of 2 equiv of LiBr to SmI_2 in THF provides $SmBr_2$,[3] and Namy reported that the mixing of 1,1,2,2-tetrabromoethane with Sm metal in THF provides ready access to $SmBr_2$.[4] Hilmersson subsequently found that microwave heating of Sm with bromine or 1,1,2,2-tetrabromoethane greatly enhances the rate of production of $SmBr_2$.[5] The main advantage of using $SmBr_2$ over SmI_2 is that it is a very efficient reagent for the pinacol homocoupling of aldehydes and ketones and, in some cases, cross coupling between different carbonyl compounds is efficient.[2] It has also been reported that $SmBr_2$ is capable of selectively reducing ketones in the presence of alkyl halides, but the synthetic applications of this selectivity have not been fully explored.[6]

Organic Synthesis Using Samarium Diiodide: A Practical Guide
By David J. Procter, Robert A. Flowers, II and Troels Skrydstrup

Published by the Royal Society of Chemistry, www.rsc.org

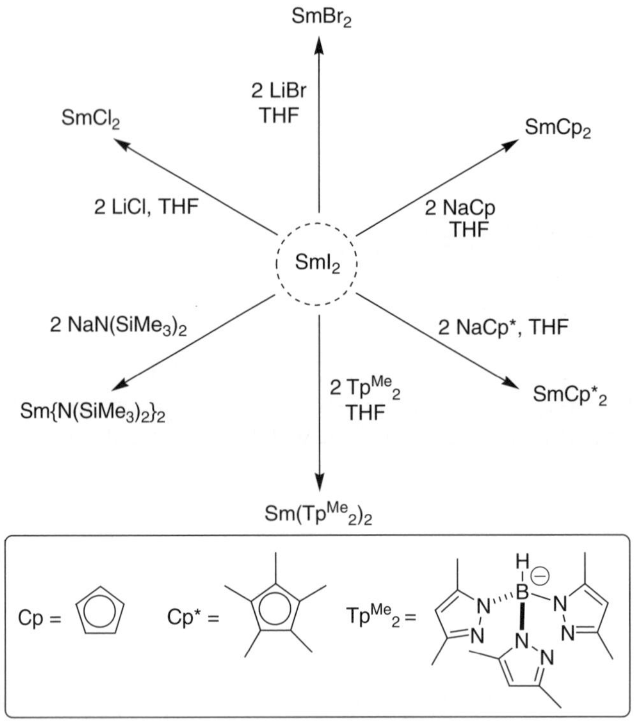

Scheme 7.1

The synthesis of $SmCl_2$ was first reported by Rossmanith in 1979 by reduction of $SmCl_3$ with a lithium metal–naphthalene dispersion in THF.[7] The preparation of $SmCl_2$ can also be carried out through the reaction of SmI_2 with 2 equiv. of LiCl in THF.[3] Surprisingly, $SmCl_2$ can also be prepared in water using a combination of Sm metal and $SmCl_3$.[8] The *in situ*-generated reagent has been used for pinacol coupling reactions. $SmCl_2$ has also been used as a critical component in the intramolecular coupling of isocyanates and α,β-unsaturated cyclic ketones to produce spirooxindoles (Scheme 7.2).[9] This method is a critical step in Wood's synthesis of the natural product (±)-welwitindolinone A isonitrile (see Section 7.4).[9]

Representative procedure.[9] The isonitrile **1** was dissolved in dry THF (10 ml) and *t*-BuOH (0.27 mmol) was added. The solution was cooled to –78 °C and degassed with nitrogen for 10 min. A mixture of SmI_2 (0.57 mmol in 5.7 ml of THF) and LiCl (2.3 mmol) was stirred for 10 min at room temperature then added over 5 min *via* cannula to the solution of **1**. After complete addition of the SmI_2–LiCl mixture, the reaction solution was stirred for 2 min at −78 °C, then quenched by the addition of saturated aqueous NH_4Cl. The mixture was allowed to warm to room temperature, then diluted with water (20 ml) and EtOAc (20 ml). The organic layer was separated and the aqueous layer was extracted with EtOAc (2 × 20 ml). The organic layer was dried over Na_2SO_4, concentrated under reduced pressure and

NCO O 1 — SmI_2, LiCl, *t*–BuOH, THF, –78 °C, 88% → O HN O 2

Scheme 7.2

purified by silica gel chromatography (gradient elution, 20–40% EtOAc–hexanes). The spirooxindole product **2** was recovered as a viscous oil in 88% yield.

7.1.2 Sm(II) Triflate

Samarium(II) triflate [$Sm(OTf)_2$] has also been utilised in place of SmI_2. For $Sm(OTf)_2$-mediated reactions, THF, CH_3CN and DME are the most commonly utilised solvents. The $Sm(OTf)_2$–THF solvate can be prepared by the reduction of $Sm(OTf)_3$ with 1 equiv of either *s*-BuLi[10] or *s*-BuMgCl.[11] The CH_3CN solvate can be prepared by the reaction of a hypervalent sulfur triflate (1,5-dithioniabicyclo[3.3.0]octane) with samarium metal and a catalytic amount of iodine in CH_3CN at 50 °C.[12] The DME solvate was prepared by reacting $Sm(OTf)_3$ and samarium metal in the presence of a catalytic amount of Hg in DME.[13] When $Sm(OTf)_2$ is generated using alkyllithium or Grignard reagents, the corresponding metal salts are present in solution and as a result the mechanistic role of the Mg or Li counterion in subsequent reactions has not been determined. Furthermore, the UV–visible spectrum of the $Sm(OTf)_2$–THF species produced from reduction of $Sm(OTf)_3$ by Grignard reagents is different from that produced through reduction by an organolithium reagent.[12] As a result, the species is reported as '$Sm(OTf)_2$'.

Although the use of $Sm(OTf)_2$ has been limited, in some cases it has been shown to be a more effective reagent than SmI_2.[14] For example, using samarium Grignard conditions (see Chapter 5, Section 5.4), $Sm(OTf)_2$ demonstrated better diastereoselectivity than SmI_2 for the addition of alkyl iodides to 2-methylcyclohexanone, 4-*tert*-butylcyclohexanone and 2-phenylpropanal. In more recent work, Fukuzawa has shown that $Sm(OTf)_2$ is superior to SmI_2 in the pinacol coupling (see Chapter 5, Section 5.1) of chiral formylferrocenes such as **3**, giving the desired (*R*,*R*)-diastereoisomer **4** predominantly in excellent yield while the use of SmI_2 gave the (*R*,*S*) isomer with significantly reduced selectivity and yield (Scheme 7.3).[15]

In this work, a range of organolithium and Grignard reagents were examined for the preparation of $Sm(OTf)_2$ from the trivalent precursor. Interestingly, the use of *n*-BuLi in place of *s*-BuLi led to a decrease in the selectivity for the (*R*,*R*)-diastereoisomer in the subsequent pinacol coupling reduction, while the use of Grignard reagents to form $Sm(OTf)_2$ led to the formation of unwanted side-products. The differences in stereoselectivity and yield upon even modest

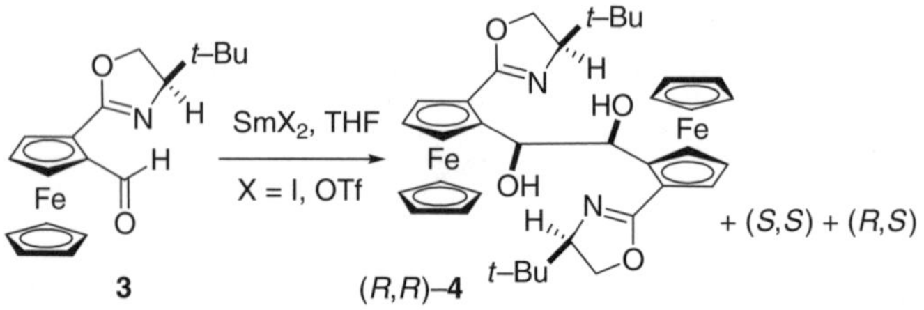

reductant	isolated yield (%)	(R, R) : (S, S) : (R, S)
SmI_2	55	9 : 44: 47
$Sm(OTf)_2$[a]	99	87: 2 : 11

[a]Produced from an equimolar combination of $Sm(OTf)_3$ and s–BuLi in THF

Scheme 7.3

changes to the organometallic reagent suggest that it is important to screen different methods for the preparation of $Sm(OTf)_2$ carefully.[15]

Representative procedure. After drying at 180 °C under vacuum, $Sm(OTf)_3$ (1.2 mmol) was placed in a flask and 5 ml of THF were added. After stirring the mixture for 1 h at room temperature, *s*-BuLi (1.0 M, 2.5 ml) in cyclohexane was slowly added to the suspension at 0 °C. The solution was then warmed to room temperature over a 1 h period, during which time a purple colour indicated the formation of $Sm(OTf)_2$. Aldehyde **3** (0.5 mmol) in 5 ml of THF was added to the $Sm(OTf)_2$ solution and stirred for 1 h. The solution was quenched with saturated NH_4Cl and extracted with Et_2O. The organic extract was washed with brine and dried over $MgSO_4$. After concentration *in vacuo*, ^1H NMR and HPLC analysis of the residues revealed the formation of three diastereoisomers, with a combined yield of 99%, of which the (*R*,*R*)-diastereoisomer **4** was the major product. The diastereoisomers were separated by silica gel chromatography [hexanes–EtOAc (4:1–2:1) eluent].

7.1.3 Sm(II) Cyclopentadienyl-based Complexes

Ligands based on the cyclopentadienyl unit have played an important role in the development of Sm(II) chemistry. Although there are many cyclopentadienyl ligands, C_5H_5 (Cp) and C_5Me_5 (Cp*) are the most commonly used ligands of the class. The synthesis of $SmCp_2$ and $SmCp^*_2$ is straightforward and most readily performed by the reaction of 1 equiv of SmI_2 with 2 equiv of the appropriate cyclopentadienylide salt in THF.[16,17] Although the insolubility of $SmCp_2$ in THF and other commonly used organic solvents has limited its use in synthesis, there are a few examples where this reagent is reported to have advantages over SmI_2. For example, the Barbier-type coupling of alkyl halides and ketones or aldehydes (see Chapter 5, Section 5.4) using $SmCp_2$ can give higher yields of products than the same reactions employing SmI_2.[16,18]

In 1981, Evans reported the synthesis of $SmCp^*_2$.[19] This reagent is soluble in a wide range of organic solvents including benzene, THF and hexanes, enabling the chemistry of Sm(II) to be investigated across a broad range of conditions. This seminal report led to the use of Sm(II) in a number of diverse areas, including inorganic synthesis, catalysis and polymerisation reactions. A complete discussion is beyond the scope of this book, but there are a number of comprehensive reviews on the topic.[20–22]

7.1.4 Sm(II) Amides, Alkoxides and Borates

Sterically congested monoanionic nitroanions, oxyanions and borates are primarily used as ligands to promote the formation of monomeric Sm(II) reductants. Among alkoxides, the 2,6-di-*tert*-butyl-4-methylphenoxide ligand provides the steric and electronic requirements necessary to form stable Sm(II) complexes. Among nitrogen-based ligands, the bis(trimethylsilyl)amide ligand has been the most utilised ligand to produce stable, soluble Sm(II) species. Although neither of these classes of reductants has been utilised extensively in organic synthesis, they have been used to study the mechanism of ketone[23] and nitro group reduction.[24] Furthermore, Flowers and Hilmersson recently reported that $[(Me_3Si)_2N]_2Sm$ in THF is useful in the reduction and reductive coupling of aldimines and ketimines.[25] Another ligand that has become useful in the study of Sm(II) is hydrotris (3,5-dimethylpyrazolyl) borate ($Tp^{Me}2$). $Sm(Tp^{Me}2)_2$ has unique reactivity compared with other Sm(II) reductants but unfortunately is insoluble in most organic solvents and, as a result, its use in organic synthesis has not been explored.[26]

7.1.5 Conclusion

SmI_2 is a valuable precursor to other Sm(II) and Sm(III) species. As discussed in Chapter 2, *in situ* modification of the reagent using additives and cosolvents is the most convenient and the most common method for generating new Sm(II) species. The synthesis and characterisation of new Sm(II) reagents, prior to their deployment in reactions, may, however, prove a key tool in the bid to develop new selective reactions and to unravel the mechanistic complexities that surround the organic chemistry of SmI_2.

7.2 Catalytic SmI_2 Reagent Systems

7.2.1 Reductive Processes Using Catalytic SmI_2

The majority of reactions using SmI_2 require a stoichiometric excess of reductant. Due to the high molecular mass of SmI_2 (MW = 404), a large amount of metal with respect to the substrate is required. Economic and environmental concerns have led to attempts to develop reagent systems that use a catalytic amount of SmI_2 and stoichiometric quantities of an inexpensive co-reductant. While a practical solution to this problem remains elusive, this section will describe the progress that has been made towards the use of sub-stoichiometric amounts of SmI_2 in organic synthesis.

A catalytic cycle using a SmI_2–Mg–TMSCl reagent system was developed by Endo for pinacol coupling (Scheme 7.4).[27] This system utilises magnesium metal as an inexpensive co-reductant to regenerate the active divalent samarium species in the catalytic cycle. Products of pinacol coupling using a range of ketone and aldehyde substrates were produced in moderate to good yields. In Endo's system, the presence of trimethylsilyl chloride is critical for the release of Sm(III) from the pinacol product, thus enabling reduction by Mg to regenerate Sm(II) in solution (Scheme 7.4).[27]

A more extensive study on the use of Mg as a co-reductant in catalytic SmI_2 pinacol couplings was carried out by Aspinall and Greeves.[28] In this study, tetraglyme was used as an additional additive and dimethyldichlorosilane was used in place of trimethylsilyl chloride (Scheme 7.5).

The use of this protocol provided good to excellent diastereoselectivities in a range of intermolecular and intramolecular pinacol coupling reactions.

Scheme 7.4

R^1	R^2	yield (%)	*rac:meso*
t–Bu	H	76	95:5
c–Hex	H	63	81:19
Ph	Me	74	94:6

Scheme 7.5

Coordination of glyme to both Sm(II) and Sm(III) is believed to be important in enhancing the diastereoselectivity of the pinacol couplings in addition to facilitating the reduction of Sm(III) to Sm(II).[28]

Representative procedure.[28] Magnesium turnings (154 mg, 6.4 mmol) were added to a Schlenk tube and stirred under argon for 1 h. A solution of SmI_2 (0.1 M solution, 4 ml, 0.4 mmol, 10 mol%) in THF, dimethyldichlorosilane (0.05 ml, 0.4 mmol) and tetraglyme (0.09 ml, 0.4 mmol) were added successively to the Schlenk tube. A combination of aldehyde (1 mmol) and dimethyldichlorosilane (0.1 ml, 0.8 mmol) in THF was added dropwise using a syringe pump at a rate such that the blue colour of the solution was maintained. After addition, the mixture was filtered and quenched with tetra-*n*-butylammonium fluoride (1.0 M in 10 ml of THF) and washed with 10 ml of brine. The solution was concentrated *in vacuo* and the crude product was extracted with diethyl ether and purified by silica gel chromatography.

Orsini reported the use of a 10 mol% SmI_2–Mg reagent system in the Reformatsky reactions of α-halocarbonyl compounds, nitriles and phosphonates with ketones, aldehydes and imines.[29]

In a seminal study on the development of catalytic SmI_2 reagent systems, Corey examined a series of metal amalgams and found that the order of reactivity in the reduction of SmI_3 to SmI_2 was Mg > Al > Zn.[30] Examination of a series of conditions showed that the combination of SmI_2, LiI and $Me_3SiOSO_2CF_3$ (TMSOTf) with Zn–Hg gave excellent results in the reductive coupling of ketones with acrylates to produce γ-butyrolactones. This catalytic SmI_2 reagent system was also found to be effective for the reduction of an epoxide and the cyclisation of a haloalkyne. The latter reaction required 20 mol% SmI_2 and elevated temperature for successful conversion to product (Scheme 7.6).[30]

One of the main issues with the use of magnesium and zinc as inexpensive co-reductants is their reactivity with alkyl halides. Also, additives such as TMSOTf are relatively expensive. In an attempt to address these limitations, Namy utilised mischmetal (La 33%, Ce 50%, Nd 12%, Pr 4%, Sm and other lanthanides 1%) as the stoichiometric reductant for the regeneration of the Sm(II).[31,32] This reagent system provides an important alternative since it does not require the use of additives and mischmetal is relatively inexpensive. This system has been utilised successfully in Barbier and Reformatsky reactions, halide reductions and pinacol couplings (Scheme 7.7).[31,32]

Studies by Namy have shown that the combinations lanthanum–SmI_2, cerium–SmI_2 or Nd–SmI_2 gave results close to those of the catalytic SmI_2–mischmetal reagent system, indicating that a number of the components of mischmetal are capable of regenerating Sm(II) from Sm(III).[32]

7.2.2 SmI_2 as a Precatalyst

There are several examples of reactions that employ a catalytic amount of SmI_2 for the *in situ* formation of a catalytic Sm(III) species. These reactions do

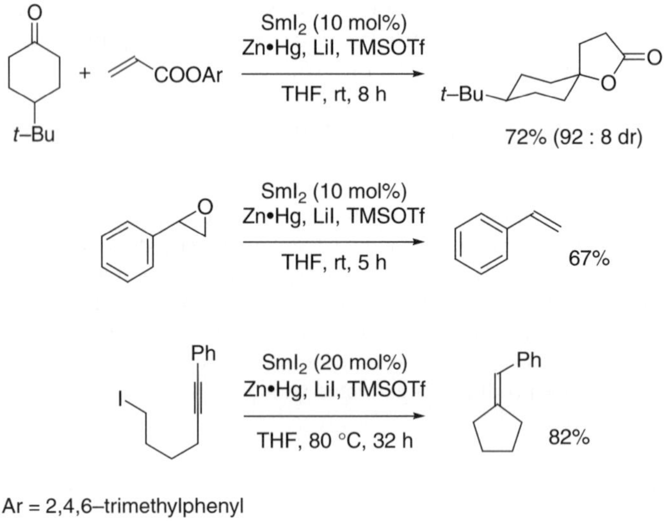

Ar = 2,4,6–trimethylphenyl

Scheme 7.6

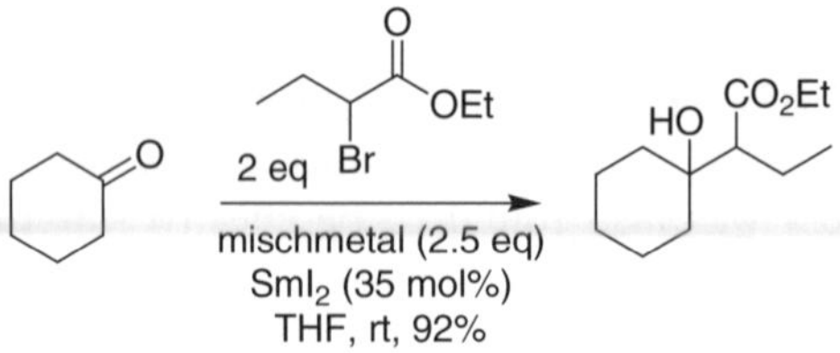

Scheme 7.7

not require a co-reductant to regenerate Sm(II). Reactions utilising substoichiometric amounts of SmI_2 as a precatalyst include the conversion of *N*-acyloxazolidinones to esters,[33] the isomerisation of α,β-epoxyamides to (*E*)-α-hydroxy-β,γ-unsaturated amides,[34] the synthesis of *N*,*N*′-disubstituted amidines[35] and α-aminophosphonates,[36] imino Diels–Alder and Mannich reactions,[37] the coupling of isocyanates with tertiary alcohols[38] and the intramolecular Tishchenko reaction.[39]

Amongst this class of reactions, the SmI_2-catalysed intramolecular Tishchenko reaction developed by Evans and Hoveyda has found widespread application.[39] In this transformation, excess aldehyde is used with a substoichiometric amount of SmI_2 to convert β-hydroxy ketones into *anti*-diol monoesters in high yield and high diastereomeric excess (Scheme 7.8).

The reaction is proposed to proceed through a six-membered Sm(III) chelate between the ketone carbonyl and the hemiacetal oxygen (Scheme 7.8).[39] Deuterium labelling studies with CD_3CDO result in complete deuterium incorporation at the newly generated carbinol centre. Since aldehydes are readily

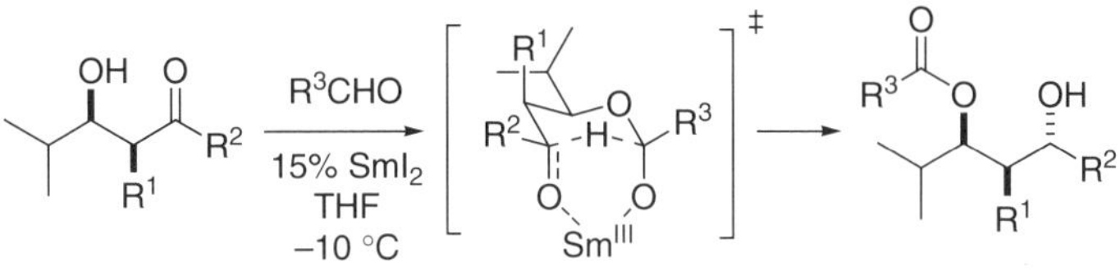

R^1	R^2	R^3	Yield (%)	*anti:syn*
H	*n*–Hexyl	Me	96	>99 : 1
H	*n*–Hexyl	Me_2CH	95	>99 : 1
H	*n*–Hexyl	Ph	94	>99 : 1
H	Me_2CH	Me	85	>99 : 1
H	Me_2CH	Ph	99	>99 : 1
Me	Me_2CH	Me	85	>99 : 1
Me	Et	Me_2CH	95	>99 : 1

Scheme 7.8

reduced by SmI_2 to pinacols, it is likely that the active catalyst is a Sm(III) species with a pinacolonate ligand derived from sacrificial aldehyde.[39]

7.2.3 Conclusion

Although the catalytic SmI_2 reagent systems described in this section represent significant advances, they are not sufficiently general to have been used extensively. For example, a particular choice of stoichiometric reductant is unlikely to be compatible with all the substrate classes that are commonly used in SmI_2 reactions. In addition, during the course of reactions with SmI_2, iodide ligands are displaced through coordination to intermediates. As a result, the Sm(III) species reduced in the catalytic cycle may produce a reductant other than SmI_2. Mechanistic studies of stoichiometric SmI_2 reactions have shown that even the replacement of iodide by bromide or chloride has a dramatic impact on the reactivity of the reductant. As a result, it is probable that more than one active reductant is present in solution, resulting in a lower degree of stereo- and chemoselectivity in most reactions involving sub-stoichiometric SmI_2. Nonetheless, as more is learned about the mechanisms of stoichiometric SmI_2 reductions, this information can be used in conjunction with the catalytic reagent systems reported to date to design an efficient, versatile and practical approach to the problem of using catalytic SmI_2.

7.3 SmI_2 in Solid-phase and Fluorous Synthesis

The use of a phase tag as a purification handle to facilitate synthesis is an important strategy for the synthetic organic chemist. In addition to the insoluble polymer supports used in 'classical' solid-phase organic synthesis, recent years have seen the emergence of soluble phase tags including perfluoroalkyl tags or *fluorous* tags. SmI_2 is emerging as a valuable tool for tackling problems in phase tag-assisted synthesis.[40]

7.3.1 Carbon–Carbon Bond-forming Processes

The first use of SmI_2 in solid-phase chemistry was reported by Armstrong in 1997.[41] An efficient synthesis of benzofuran derivatives on a solid support was achieved by SmI_2-mediated cyclisation of unsaturated aryl iodides such as **5** and **6**. Benzofuran products were obtained in good yield after cleavage of the products from the support. Armstrong also carried out sequential processes where the intermediate radicals from the cyclisation of iodides, such as **7**, were reduced by a second equivalent of SmI_2 and the resultant organosamarium was captured by a carbon electrophile to give adducts **8** after hydrolytic cleavage of the products from the support (Scheme 7.9).[41,42] Armstrong's pioneering work was crucial in illustrating the compatibility of SmI_2 with the common classes of polymer support used in solid-phase synthesis.

Linhardt employed Reformatsky-type additions to carbon electrophiles (see Chapter 5, Section 5.5) in a solid-phase synthesis of *C*-sialosides.[43] Immobilised sialyl donor **9** was treated with SmI_2 in the presence of ketone and aldehyde

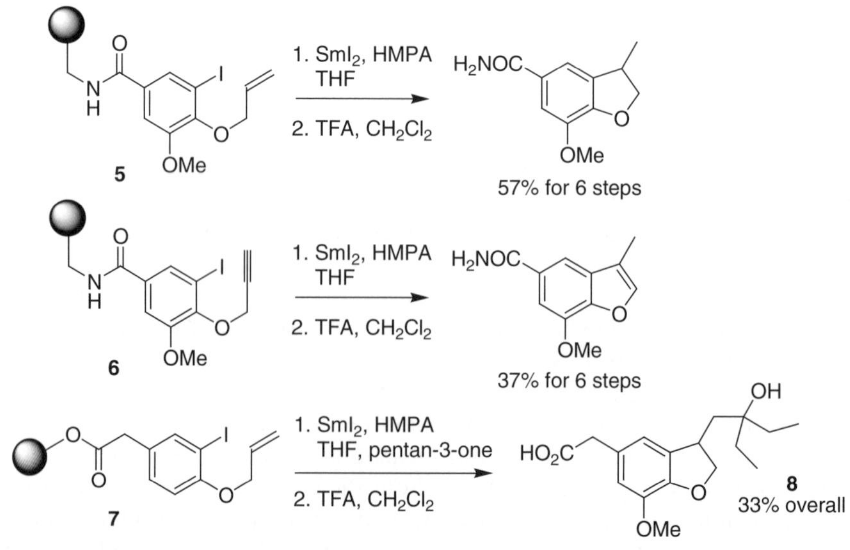

Scheme 7.9

Scheme 7.10

electrophiles to give adducts such as **10**. Cleavage from the support gave *C*-glycoside **11** in good overall yield (Scheme 7.10).[43]

Procter reported a solid-phase variant of Fukuzawa's asymmetric γ-butyrolactone synthesis (see Chapter 5, Section 5.2) that involves the intermolecular reductive coupling of aldehydes and ketones with α,β-unsaturated esters, immobilised using an ephedrine chiral resin **12**.[44] For example, treatment of acrylate **13** and crotonate **14** with cyclohexanecarboxaldehyde, employing SmI_2 in THF, with *tert*-butanol as a proton source, gave lactones **15** and **16**, respectively, in moderate yield and good to high enantiomeric excess (Scheme 7.11).[44] The ephedrine resin **12** can be conveniently recovered and recycled.[45]

Representative procedure. A suspension of (1*R*,2*S*)-*N*-Wang-bound ephedrinyl crotonate resin **14** (0.49 mmol, 3.2 equiv) in THF (5 ml) under argon was gently stirred for 15 min prior to the addition of cyclohexanecarboxaldehyde (18.4 ml, 0.15 mmol, 1 equiv) and *tert*-BuOH (2 equiv). The resultant suspension was then allowed to stir for another 60 min at room temperature before being cooled to –15 °C. A precooled SmI_2 solution (0.1 M in THF, 5.5 equiv) was then added and the dark blue solution was allowed to stir at –15 °C until TLC analysis showed that the aldehyde had been consumed. The reaction mixture was then allowed to warm to room temperature over 5–6 h. The resin was removed by filtration and washed with THF. The filtrate was concentrated (to ~30 ml) and washed with aqueous saturated NaCl (4 ml). The aqueous layer was separated and washed with Et_2O (3 × 15 ml). The combined organic layers were then dried (Na_2SO_4) and concentrated *in vacuo*. The crude product was purified by column chromatography on silica [eluting with 10% EtOAc–petroleum ether (b.p. 40–60 °C)] to give **16** (18.2 mg, 66%) as a pale yellow solid.

Procter developed the new, fluorous-tagged chiral auxiliary **17** for the asymmetric, SmI_2-mediated coupling of aldehydes and α,β-unsaturated esters.[46] γ-Butyrolactones are obtained in moderate to good isolated yield and in high enantiomeric excess. The fluorous tag allows the auxiliary to be conveniently recovered by fluorous solid-phase extraction (FSPE) and reused (Scheme 7.12).

13 R = H
14 R = Me

SmI_2, THF
t-BuOH, −15 °C
R = H

15
66% (76% ee)

SmI_2, THF
t–BuOH, −15 °C
R = Me

16
66% (96% ee)

12

Scheme 7.11

SmI_2
t-BuOH
THF, −15 °C

17

75%
98% ee
(only diastereoisomer)

FSPE for recovery and re–use

Scheme 7.12

7.3.2 Linker Systems Cleaved Using SmI_2

The mild, neutral electron-transfer conditions associated with the use of SmI_2 make the reagent ideal for the selective cleavage of linkers in solid-phase organic synthesis and other types of phase tag-assisted synthesis.

The first linker systems designed to be cleaved using SmI_2 were based on the reduction of N–O bonds. For example, Abell prepared amides (including **19**) and ureas using an N–O linker cleaved by SmI_2,[47] and Taddei used a similar strategy in a solid-phase synthesis of β-lactams such as **20** (Scheme 7.13).[48]

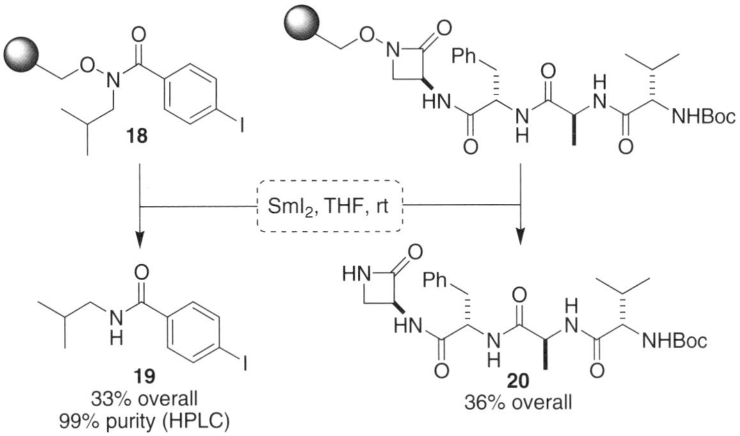

Scheme 7.13

Representative procedure. The resin **18** (1.21 g, 1.05 mmol) was pre-swollen with THF (2.1 ml) and SmI_2 (0.1 M in THF, 21.0 ml, 2.10 mmol) was added. The reaction suspension was shaken at 25 °C for 3 h. The resin was filtered off and rinsed with CH_2Cl_2 (5 × 10 ml) and the cleavage solution and washings were collected. The filtrate was evaporated to give a dark yellow residue which was redissolved in a solution of Et_2O (25 ml), 1 M HCl (20 ml) and 10% aqueous sodium thiosulfate (5 ml). The mixture was transferred to a separating funnel and shaken until it became colourless. The organic layer was collected and the aqueous layer was extracted with Et_2O (2 × 20 ml). The combined organic layers were washed with brine (2 × 20 ml) and dried ($MgSo_4$). The solid obtained after evaporation was redissolved in the minimum amount of CH_2Cl_2 (~0.3–0.5 ml) and filtered through a short pad of silica (eluting with 20% ethyl acetate in hexanes). The filtrate was collected and evaporated to afford 4-iodo-*N*-isobutylbenzamide **19** (49 mg, 33%).

Procter developed a new class of linker for the synthesis of carbonyl compounds that is cleaved using SmI_2. The cleavage of the linker is based on the reduction of α heteroatom-substituted carbonyl compounds to the parent ketones, esters or amides using SmI_2 (see Chapter 4, Section 4.5). Oxygen[49] and sulfur-based linkers[50] have been used to synthesise a range of products (Scheme 7.14). Procter has also shown the linker to be effective in fluorous synthesis,[51] where the polymer phase tag is replaced by a perfluoroalkyl group or fluorous tag. In some cases, DMPU and LiCl are used as additives to increase the reduction potential of the reagent and to facilitate the reduction (see Chapter 2, Section 2.3).

Representative procedure. DMPU (967 μl, 8.00 mmol) was added to a suspension of **21** (0.498 mmol) in THF (3 ml). SmI_2 (0.1 M in THF, 25.0 ml, 2.50 mmol) was added dropwise and the mixture stirred slowly for 12 h. The resin was then removed by filtration and washed with THF (100 ml). The filtrate was concentrated *in vacuo* to give the crude product as a yellow oil. Filtration through a

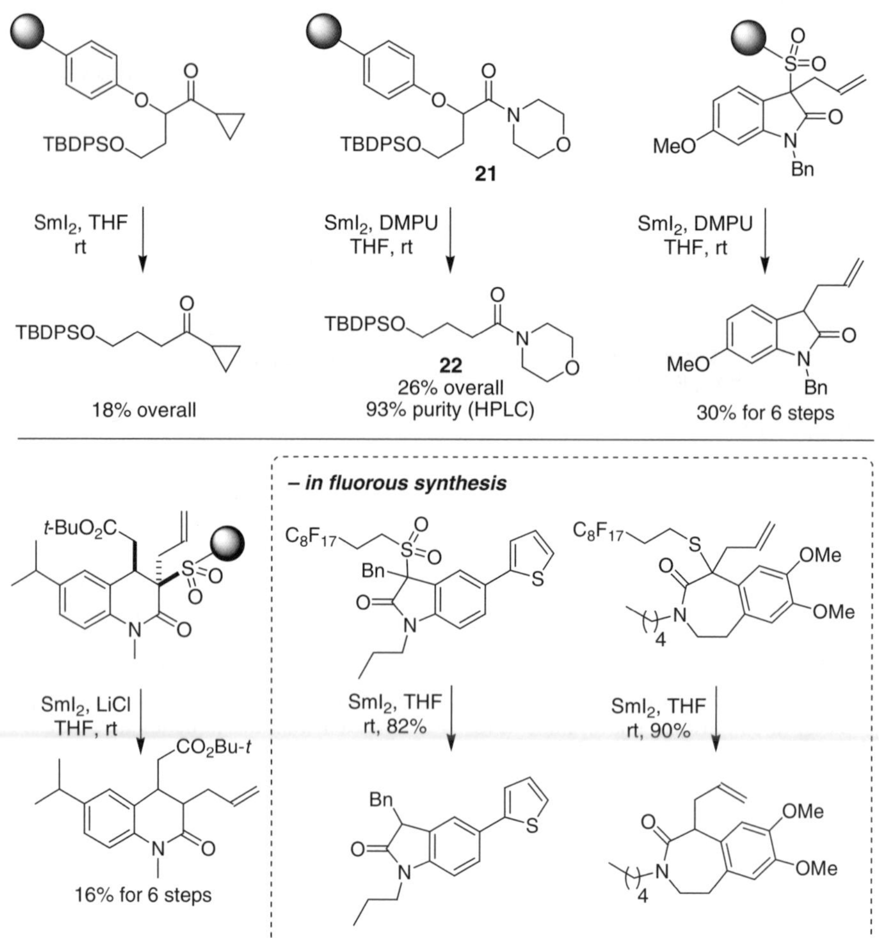

Scheme 7.14

short pad of silica gel (eluting with 50% EtOAc–petroleum ether) gave **22** (54 mg, 0.131 mmol, 26% for four steps from the phenol resin starting material).

Cleavage of the linker with SmI_2 can also be used to trigger cyclisation reactions; for example, treatment of immobilised sulfone **23** with SmI_2 results in cleavage of the linker and cyclisation to give tetrahydroquinoline **24**. This cyclisation may involve either a radical or samarium enolate addition to the electron-deficient alkene (Scheme 7.15).[50] In the cleavage of the linker in **25**, it was found that varying the order of addition of SmI_2 gave two very different cyclisation products: slow addition of SmI_2 to the substrate gave spirocycle **26** resulting from alkylation of an intermediate samarium enolate, whereas slow addition of the substrate to SmI_2, gave indolocarbazole **27**, which appears to

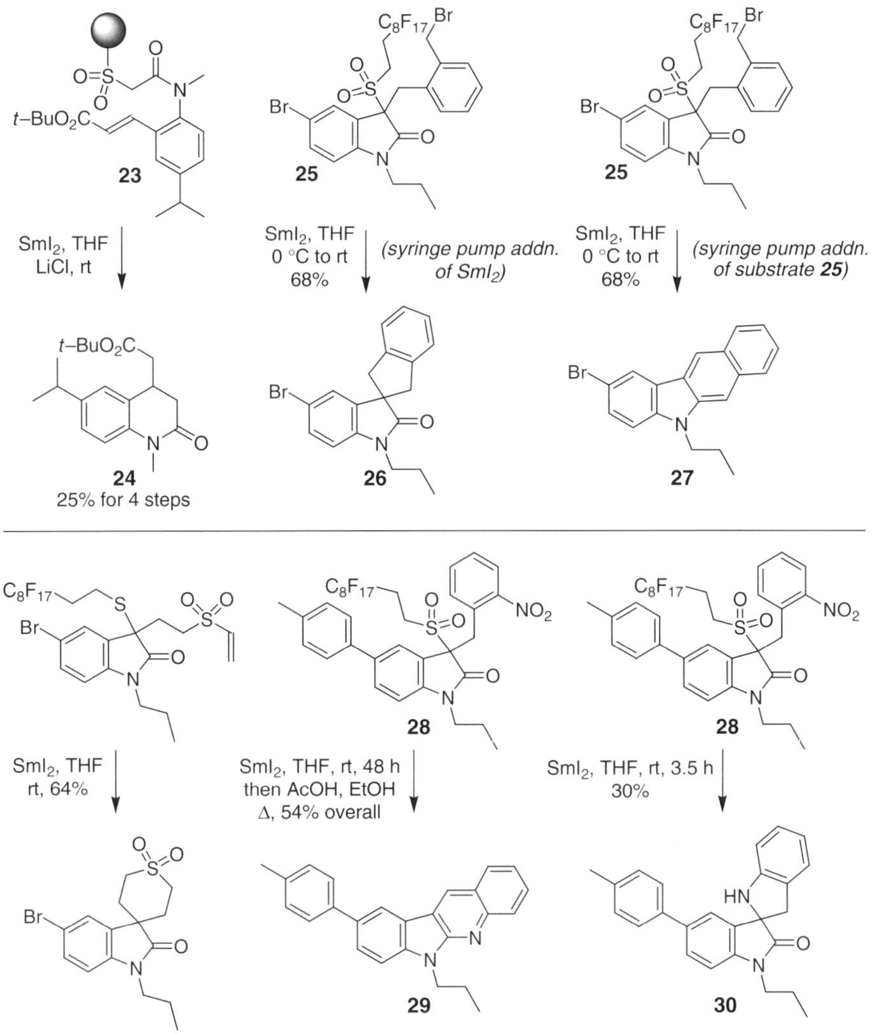

Scheme 7.15

result from the Barbier cyclisation of a benzylic samarium intermediate (Scheme 7.15).[52] Carbon–heteroatom bonds can also be formed in cleavage–cyclisation processes. For example, treatment of **28** with SmI_2 over 48 h results in sequential removal of the fluorous tag and reduction of the aryl nitro group. Subsequent acid-mediated cyclisation of the intermediate aniline gives indoloquinoline **29** in moderate overall yield (Scheme 7.15).[53] This approach has been used in a recent fluorous synthesis of the indoloquinoline natural product cryptolepine.[54] Interestingly, interrupting the SmI_2 reduction of **28** after 3.5 h gave azaspirocycle **30** arising from attack of the samarium enolate intermediate at the nitrogen of the nitro group (Scheme 7.15).[53]

7.3.3 Conclusion

Although the use of SmI_2 in phase tag-assisted synthesis is an area still in its infancy, successfully harnessing the synthetic power of SmI_2 in reactions with immobilised or tagged substrates promises benefits for library synthesis, asymmetric methodology and the design of new linker systems. With the compatibility of the reagent now well established with a variety of polymer supports and an ever growing number of examples of reactions typical of the reagent involving immobilised or tagged substrates, the way is clear to bring the full potential of SmI_2 to phase tag-assisted synthesis.

7.4 SmI_2-mediated Cyclisations in Natural Product Synthesis

The total synthesis of natural products provides a rigorous testing ground for new reagents and new reactions. Many natural product targets have complex structures rich in functionality and sequences of highly selective reactions are required for their synthesis. The mild and selective nature of SmI_2 has made it an important tool for natural product synthesis and, in particular, the reagent has been used to mediate cyclisations to form a wide variety of carbocyclic and heterocyclic systems. As Procter comprehensively reviewed the use of SmI_2-mediated cyclisations in natural product synthesis in 2004,[55] this section provides a selection of 'classical' applications of SmI_2-mediated cyclisations, alongside 'contemporary' examples from the recent literature. The examples are organised according to the size of the ring formed in the SmI_2-mediated cyclisation.

7.4.1 Four-membered Ring Formation Using SmI_2

7.4.1.1 A Synthesis of Paeoniflorin

In 1993, Corey reported the first synthesis of paeoniflorin.[56] The core of paeoniflorin was constructed using a SmI_2-mediated Reformatsky-type reaction (Scheme 7.16). Treatment of α-chloronitrile **31** with SmI_2 gave cyclobutanol **32** in excellent yield. The sensitivity of cyclobutanol **32** to base precluded the use of more conventional aldol-type cyclisations.[56]

7 equiv SmI_2, THF, −45 °C, 10 h, 93 %

31 **32** paeoniflorin

Scheme 7.16

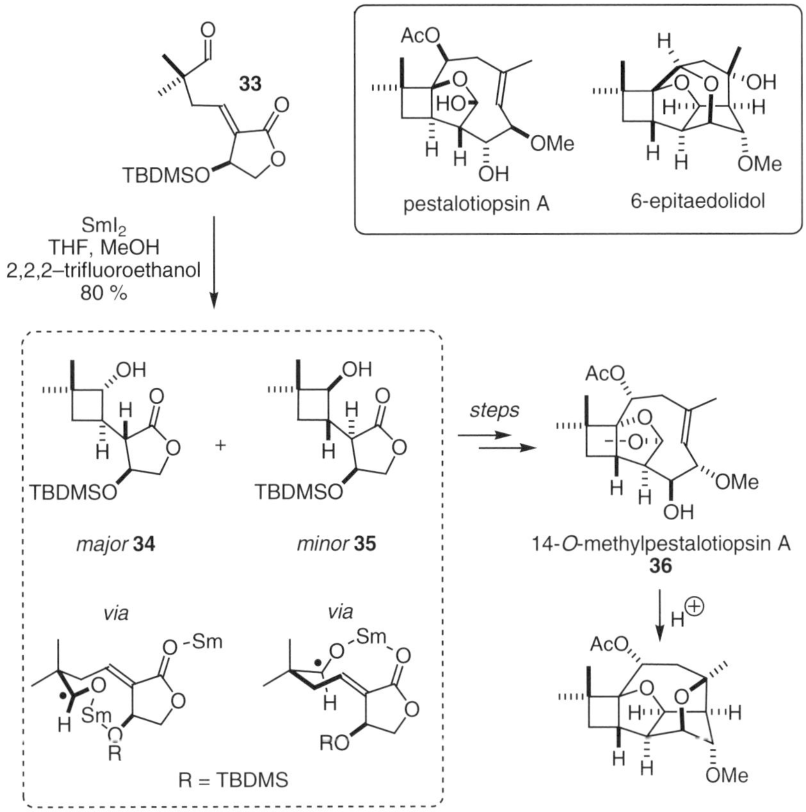

Scheme 7.17

7.4.1.2 An Approach to the Pestalotiopsin and Taedolidol Skeletons

Pestalotiopsin A and 6-epitaedolidol are structurally related caryophyllene-type sesquiterpenes. In 2003, Procter reported the use of a SmI_2-mediated 4-*exo-trig* carbonyl–alkene cyclisation to construct the core of pestalotiopsin A.[57] Treatment of cyclisation substrate **33** with SmI_2 in THF, MeOH and 2,2,2-trifluoroethanol gave cyclobutanol products **34** and **35** in good yield and with moderate diastereoselectivity. The major diastereoisomer is believed to arise from a cyclisation in which coordination to the silyl ether group directs addition of the ketyl radical anion to the alkene (Scheme 7.17).[57]

In the absence of 2,2,2-trifluoroethanol, low yields of cyclobutanols were obtained. It is thought that the acidic proton donor prevents elimination of the silyloxy group by rapid quenching of the samarium enolate **37** formed in the cyclisation (Figure 7.1).[57]

Procter has recently found that the pestalotiopsin skeleton **36** can be converted to the previously unexplored taedolidol natural products upon treatment with acid, suggesting a biosynthetic link between the two families.[58]

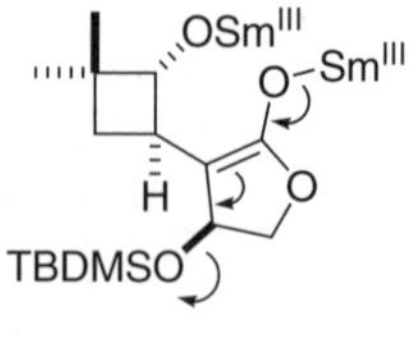

Figure 7.1 Structure of the samarium enolate **37**.

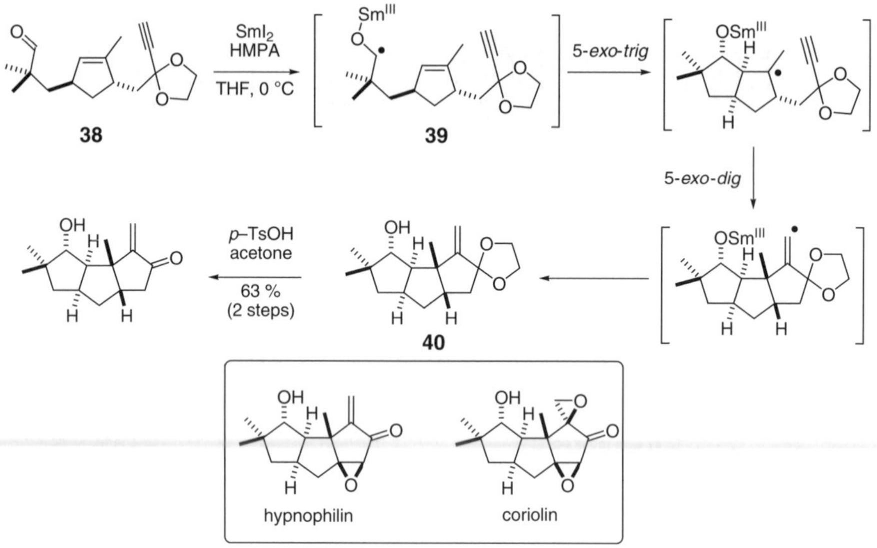

Scheme 7.18

7.4.2 Five-membered Ring Formation Using SmI_2

7.4.2.1 The Synthesis of (±)-Hypnophilin and the Formal Synthesis of (±)-Coriolin

In 1988, Curran used a SmI_2-mediated sequential process in the synthesis of (±)-hypnophilin and the formal synthesis of (±)-coriolin.[59] Treatment of aldehyde **38** with SmI_2–HMPA gave tricyclic cyclopentanol **40** (Scheme 7.18). Three new stereocentres were generated with complete diastereocontrol during the sequential cyclisations. Attempts to mediate the cyclisation of **38** using Zn–TMSCl or using photolysis led to recovery of the starting material. The use of DMPU as an alternative additive in conjunction with SmI_2 led to a decrease in the diastereoselectivity of the sequence. The reaction sequence involves 5-*exo-trig* cyclisation of the ketyl radical **39** followed by 5-*exo-dig* cyclisation of the resultant radical.[59]

7.4.2.2 A Synthesis of Grayanotoxin III

In 1994, Matsuda reported a remarkable route to grayanotoxin III in which three of the four rings of the target are formed through the use of SmI_2.[60] In the first SmI_2-mediated step, to form the CD rings of the natural product, the hemiketal **41**, which is in equilibrium with the hydroxy ketone **42**, was treated with SmI_2 in the presence of HMPA (Scheme 7.19). Addition of the ketyl radical anion to the unactivated olefin proceeded well to give **43** in high yield.[60]

In the second SmI_2-mediated cyclisation in Matsuda's approach, an allylic sulfide was used as a radical acceptor in a cyclisation to form the A ring of the natural product (Scheme 7.20).[60] The cyclisation of **44** resulted in elimination of the sulfide to generate a new double bond, which was then epoxidised and reduced to form the tertiary alcohol found in the B ring of the target. The cyclisation proceeded to form exclusively the *syn* cyclopentanol **45** in good yield. It is interesting that the stereochemistry and yield of the cyclisation are independent of the initial alkene geometry of the allyl sulfide. The seven-membered ring of the target was formed using a SmI_2-mediated pinacol cyclisation.[60]

7.4.2.3 A Synthesis and Structural Revision of (±)-Laurentristich-4-ol

In 2008, Wang and Li reported the synthesis of both the proposed and the revised structure of (±)-laurentristich-4-ol using an intramolecular SmI_2-mediated ketyl radical addition to a benzofuran to construct the spirocyclic

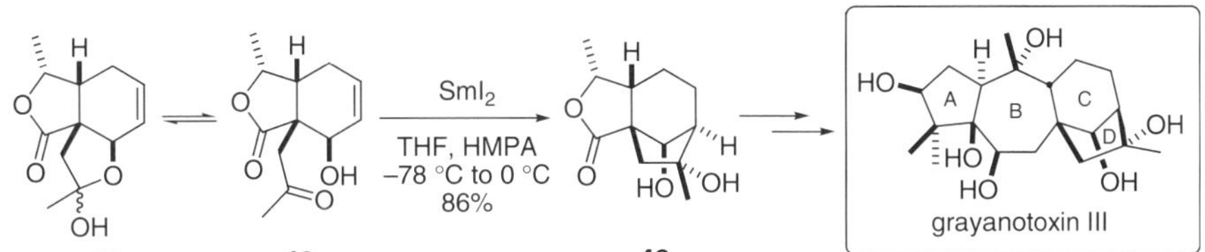

41 42 SmI_2 THF, HMPA −78 °C to 0 °C 86% 43 grayanotoxin III

Scheme 7.19

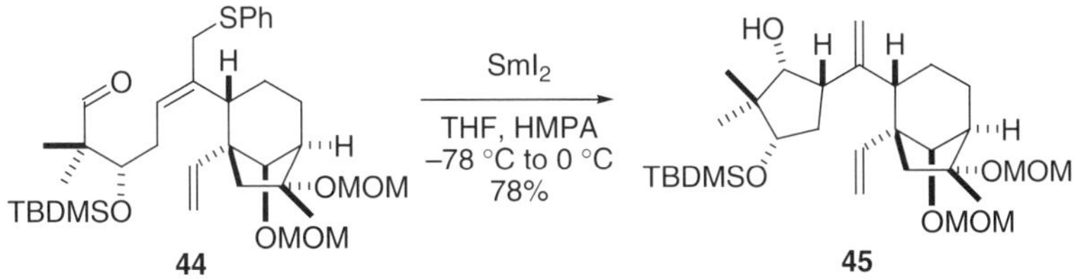

44 SmI_2 THF, HMPA −78 °C to 0 °C 78% 45

Scheme 7.20

core of the target.[61] Treatment of **46** with SmI_2–HMPA in THF gave spirocycle **47** as a single diastereoisomer in good yield (Scheme 7.21). The proposed structure of the natural product was synthesised from **47** in a further three steps. Modification of the approach allowed an isomer to be prepared, thus allowing the structure of the natural product to be revised.[61]

7.4.2.4 An Approach to (±)-Welwitindolinone A Isonitrile

In 2008, Wood used SmI_2 to mediate cyclisations in several approaches towards (±)-welwitindolinone A isonitrile.[9] The spirooxindole core of the natural product was prepared using the SmI_2-mediated intramolecular couplings of α,β-unsaturated ketones with isocyanates (Scheme 7.22). For example, treatment of isocyanate **48** with SmI_2 and LiCl gave **49** as a single diastereoisomer. The cyclisation is thought to proceed by initial reduction of the enone followed by anionic addition to the isocyanate.[9]

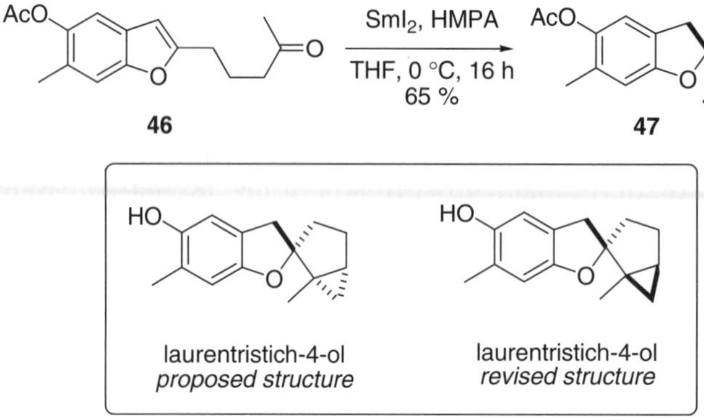

Scheme 7.21

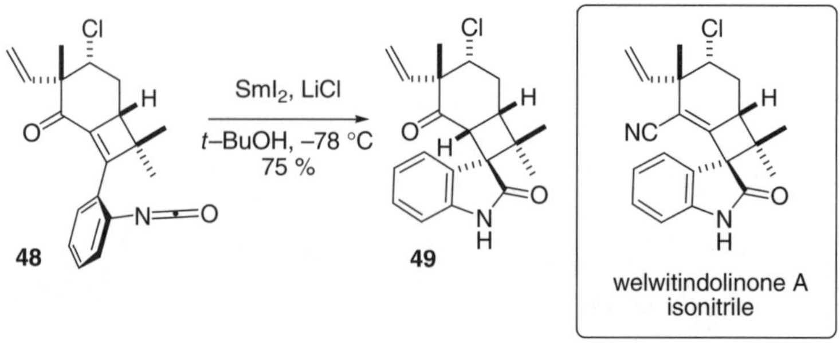

Scheme 7.22

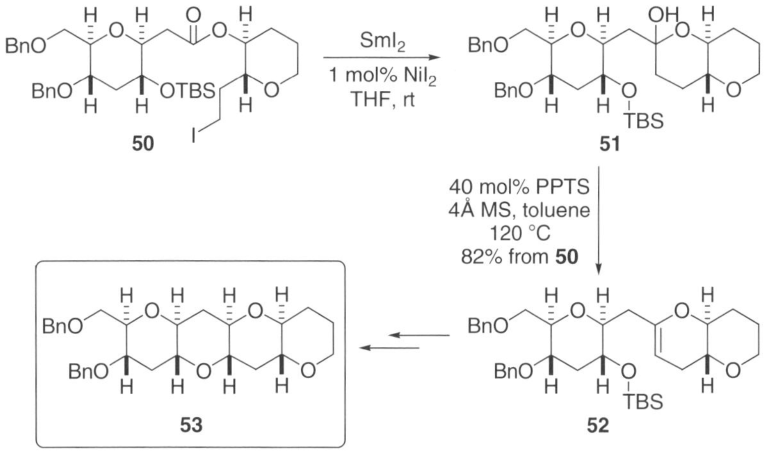

Scheme 7.23

7.4.3 Six-membered Ring Formation Using SmI_2

7.4.3.1 An Approach to Marine Polycyclic Ethers

Nakata utilised the SmI_2-mediated Barbier-type cyclisation of a primary iodide with an ester as part of a convergent synthesis of *trans*-fused 6,6,6,6-tetracyclic ethers, which are typically found in marine polycyclic ethers.[62] Treatment of iodide **50** with excess SmI_2 in the presence of 1 mol% of NiI_2 led to the smooth formation of intermediate hemiacetal **51**, which was dehydrated to give dihydropyran **52** that was then further elaborated to give tetracycle **53** (Scheme 7.23).[62]

7.4.3.2 A Synthesis of Pradimicinone

In 1999, Suzuki completed the synthesis of pradimicinone,[63] the aromatic pentacyclic aglycone moiety common to both the pradimicin and benanomicin antibiotic classes. The key step in the approach involves a SmI_2-mediated pinacol cyclisation of an axially chiral 2,2′-biaryldicarbaldehyde **54** (Scheme 7.24). This cyclisation proceeds to give the *trans*-1,2-diol **55** in quantitative yield and with complete transfer of the axial chirality in the starting material to the central chirality of the product, which is obtained in enantiomerically pure form. The selectivity and chiral transfer were attributed to an *Re*,*Re*-cyclisation mode, giving the diequatorial product.[63]

7.4.3.3 A Synthesis of (+)-Microcladallene B

In 2007, Kim reported the first asymmetric synthesis of (+)-microcladallene B.[64] In this approach, the tetrahydropyran ring of the target was prepared by a SmI_2-mediated 6-*exo-trig* carbonyl–alkene cyclisation (Scheme 7.25). The high diastereoselectivity of the cyclisation can be rationalised by invoking coordination of

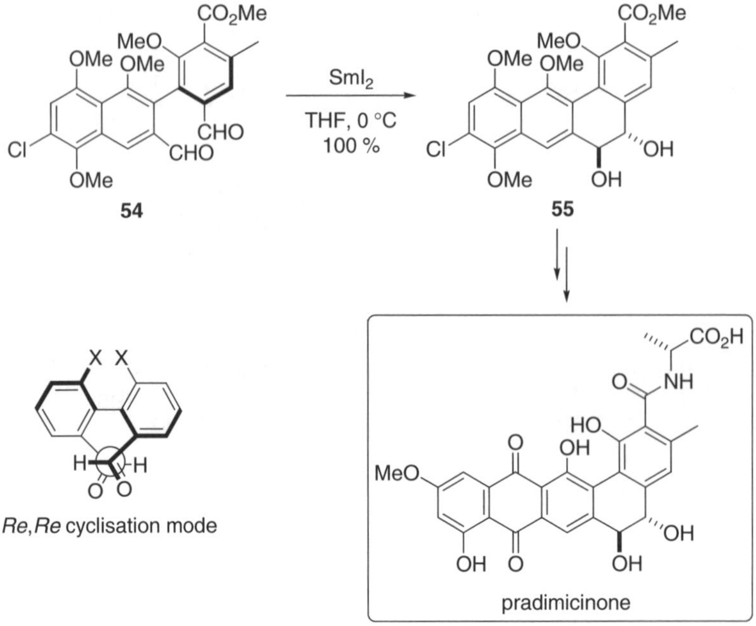

Scheme 7.24

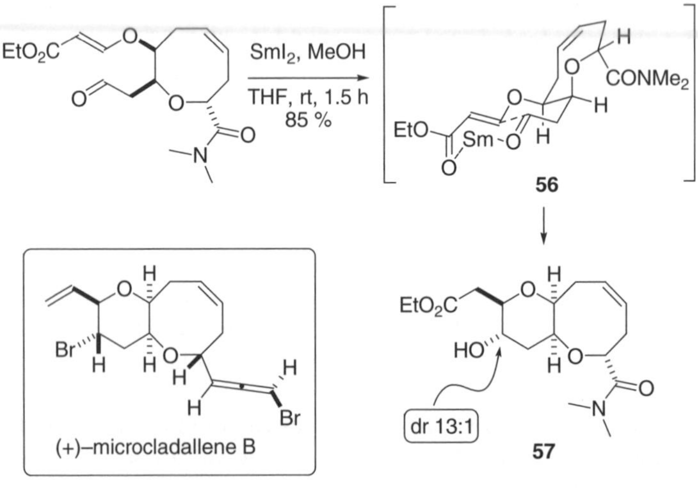

Scheme 7.25

the aldehyde and electron-deficient alkene to samarium and formation of complex **56**, in which the aldehyde adopts a pseudoequatorial orientation. Coordination of SmI_2 to the aldehyde facilitates electron transfer and additional coordination to the alkene radical acceptor facilitates cyclisation of the resultant radical. The synthesis of (+)-microcladallene B was completed in 10 steps from **57**.[64]

7.4.3.4 A Synthesis of Botcinins C, D and F

In 2008, Shiina reported the first stereoselective synthesis of botcinins C, D and F.[65] The botcinins contain a bicyclic heterocyclic core in which all of the carbon centres, apart from the lactone carbonyls, are stereogenic (Figure 7.2).

The bicyclic core of the targets was constructed using a SmI_2-mediated Reformatsky reaction (Scheme 7.26). Shiina found that the use of HMPA in the Reformatsky reaction had a dramatic effect on its outcome. When α-bromo ester **58** was treated with SmI_2 in the absence of HMPA, alcohol **60** (containing the incorrect stereochemistry required for the natural products) was obtained

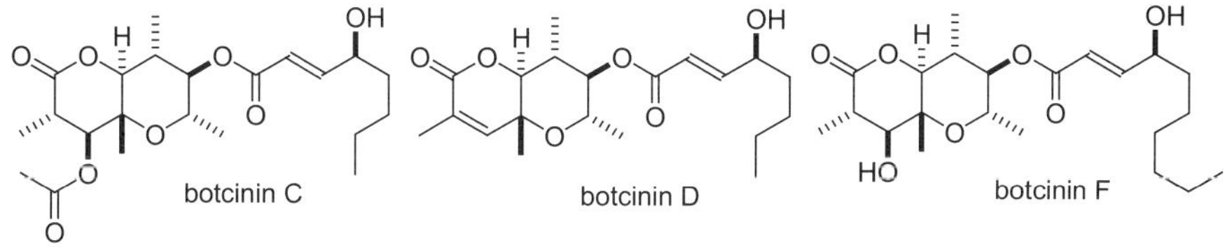

Figure 7.2 Structures of botcinins.

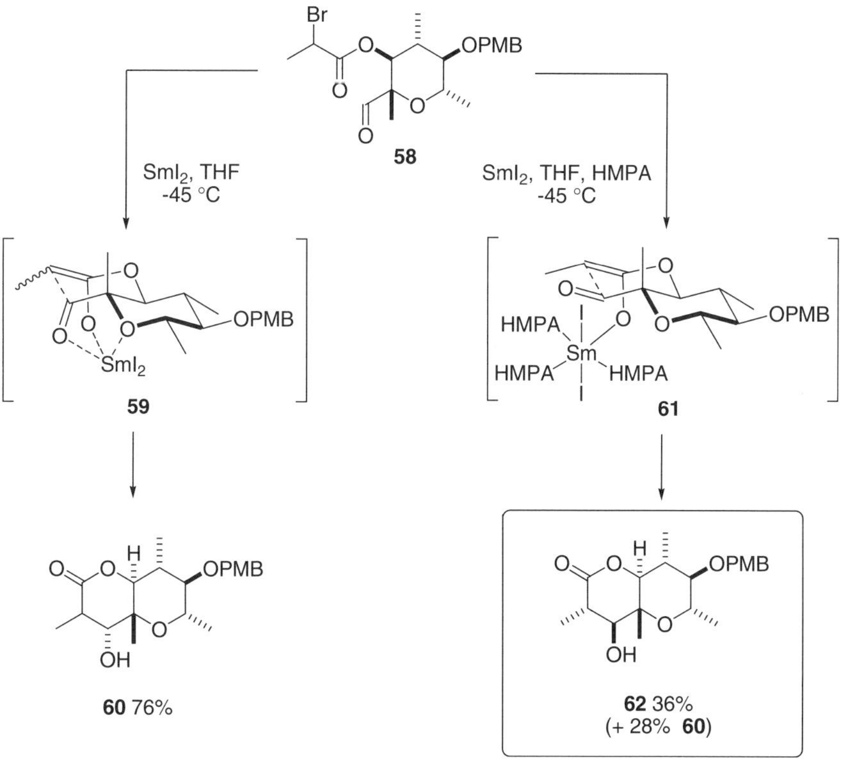

Scheme 7.26

as a mixture of diastereoisomers; **60** is thought to arise from transition structure **59** in which three oxygens in the substrate chelate to the samarium metal (Scheme 7.26).[65]

When HMPA was included in the reaction, **62**, containing the correct stereochemistry for the target, was the major product. Compound **62** is thought to arise from transition structure **61**, in which HMPA reduces the chelation of samarium to the substrate (Scheme 7.26). Lactone **62** was used to prepare all three botcinins.[65]

7.4.4 Seven-membered Ring Formation Using SmI_2

7.4.4.1 Syntheses of (–)-Balanol

In 1998, Naito reported the use of SmI_2 to form the hexahydroazepine ring of (–)-balanol through a carbonyl–oxime coupling (see Chapter 5, Section 5.1) (Scheme 7.27).[66] Attempts to form this ring using *n*-Bu_3SnH gave only a moderate yield of **64** and relatively low diastereoselectivity was observed (dr 3:1). The use of SmI_2–HMPA to mediate the cyclisation was found to give an improved yield of **64**, with higher diastereoselectivity (dr 7:1). The stereoselectivity of the cyclisation was explained by invoking *anti*-transition structure **63**. The addition of HMPA was found to be essential as no reaction was observed in its absence.[66]

In 2000, Skrydstrup reported a synthesis of balanol using a similar strategy to construct the seven-membered ring: SmI_2-mediated carbonyl–hydrazone coupling of **65** gave **66** in good yield and with high diastereoselectivity (dr 10:1) (Scheme 7.28).[67]

The addition of HMPA was found to be crucial for efficient cyclisation, as intermolecular pinacol coupling proved to be the major pathway when the aldehyde hydrazone was subjected to SmI_2 alone. Skrydstrup proposed that a

HO2C O OH NH OH (–)–balanol

SmI2 HMPA, t–BuOH THF, –78 °C to rt 12 h, 53 % OBn N Boc BocN O SmIII(HMPA)n 63 OH NHOBn Boc 64 major dr 7:1

Scheme 7.27

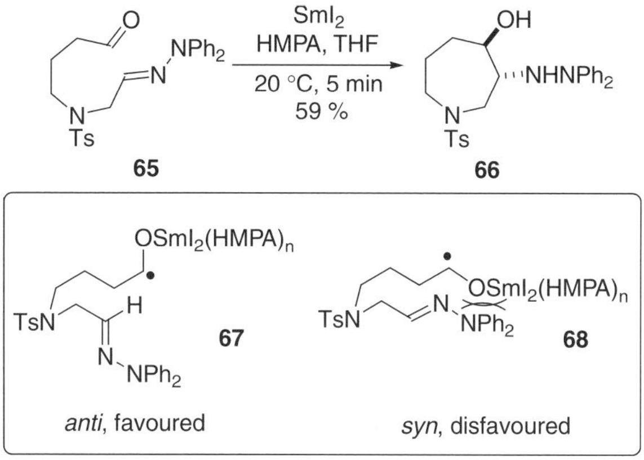

Scheme 7.28

chelated intermediate was not operating for the cyclisations with hydrazones and that simple steric interactions between the HMPA–SmI_2 bound ketyl radical anion and the diphenylhydrazone unit in transition structures **67** and **68** is the dominant factor for the *trans* selectivity observed (Scheme 7.28).[67]

7.4.5 Eight-membered Ring Formation Using SmI_2

7.4.5.1 A Synthesis of Paclitaxel (Taxol)

Mukaiyama's 1999 approach to the anti-cancer natural product paclitaxel (Taxol) involved construction of the B ring using a SmI_2-mediated Reformatsky reaction.[68] Treatment of **69** with SmI_2 at –78 °C gave the product **70** in high yield and with good diastereoselectivity (Scheme 7.29).

7.4.5.2 A Synthesis of (+)-Isoschizandrin

In 2003, Molander reported the synthesis of (+)-isoschizandrin using the SmI_2-mediated 8-*endo-trig* carbonyl–alkene cyclisation of ketone **71** (Scheme 7.30).[69] The axial chirality of the biaryl system efficiently controls the central chirality of the product. The (*Z*)-alkene geometry is also vital to the stereochemical outcome and the presence of HMPA in the reaction mixture helps control the conformation of the transition state by increasing the steric demands of the alkoxysamarium substituent.[69]

7.4.6 Nine-membered Ring Formation Using SmI_2

7.4.6.1 An Approach to Ciguatoxin

In 1998, Tachibana used a SmI_2-mediated Reformatsky reaction to form the nine-membered oxonone F ring of ciguatoxin.[70] Treatment of bromide **72** with

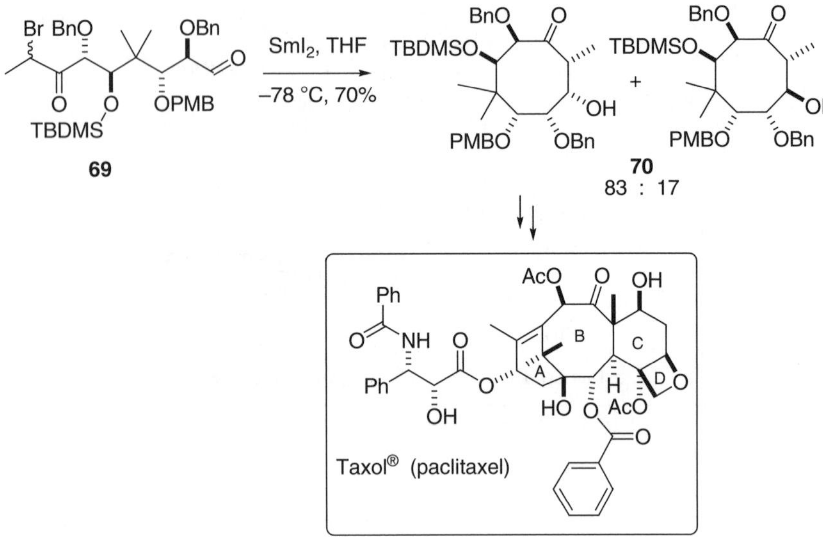

Scheme 7.29

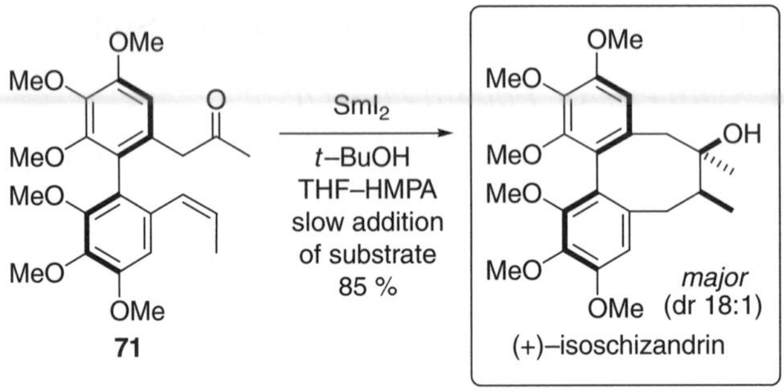

Scheme 7.30

SmI_2 gave **73** as a single diastereoisomer after *in situ* acetylation (Scheme 7.31). The product was converted to decacyclic polyether **74**, representing a model of the F–M rings of ciguatoxin.[70]

7.4.7 Forming Larger Rings Using SmI_2

7.4.7.1 A Synthesis of Diazonamide A

In 2003, Nicolaou reported a synthesis of diazonamide A that utilised a SmI_2-mediated heteropinacol coupling to construct the 12-membered macrocyclic

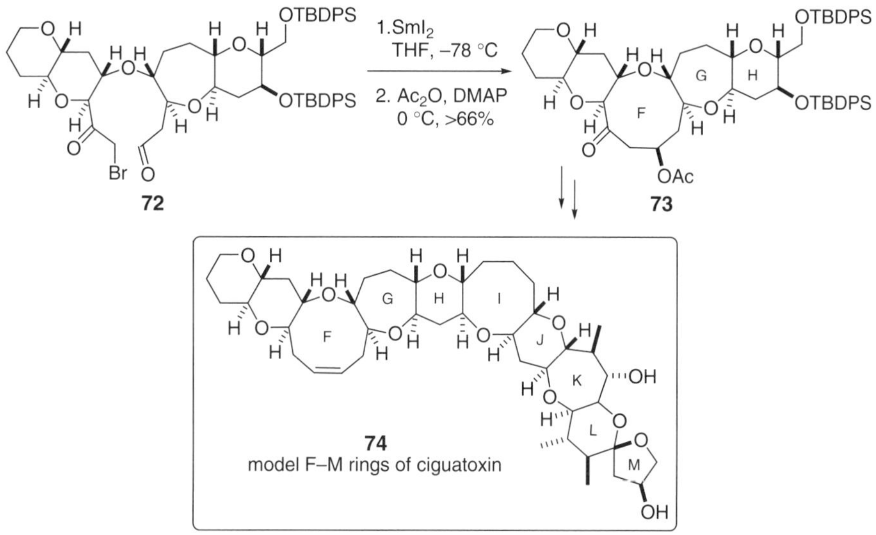

Scheme 7.31

core of the target (see Chapter 5, Section 5.1).[71] Treatment of **75** with SmI_2 in the presence of DMA led to sequential pinacol coupling and N–O bond cleavage. The crude product **76** was coupled directly with Fmoc-protected valine to give **77** in good overall yield for the three-step sequence. A diradical mechanism was proposed for the pivotal coupling step (Scheme 7.32).[71]

7.4.7.2 A Synthesis of β-Araneosene

In 2005, Corey described a synthesis of β-araneosene that exploited a SmI_2-mediated pinacol-type macrocyclisation followed by a ring expansion to build the bicyclic framework of the target.[72] Slow addition of substrate **78** to a solution of SmI_2 at reflux gave 12-membered, *anti*-pinacol product **79** as a single diastereoisomer (Scheme 7.33). Attempts to use low-valent Ti-based reagents to carry out the reductive coupling led to a mixture of uncyclised reduction products.[72]

7.4.7.3 A Synthesis of Kendomycin

In 2008, Panek reported an asymmetric synthesis of kendomycin that utilised a SmI_2-mediated Barbier-type cyclisation to form a 16-membered ring (Scheme 7.34).[73] Treatment of benzylic bromide **80** with SmI_2 gave a secondary alcohol as a single diastereoisomer. The stereochemistry of the product was not determined. Deprotection of the product and oxidation gave orthoquinone **81**, which was converted to kendomycin in two steps.[73]

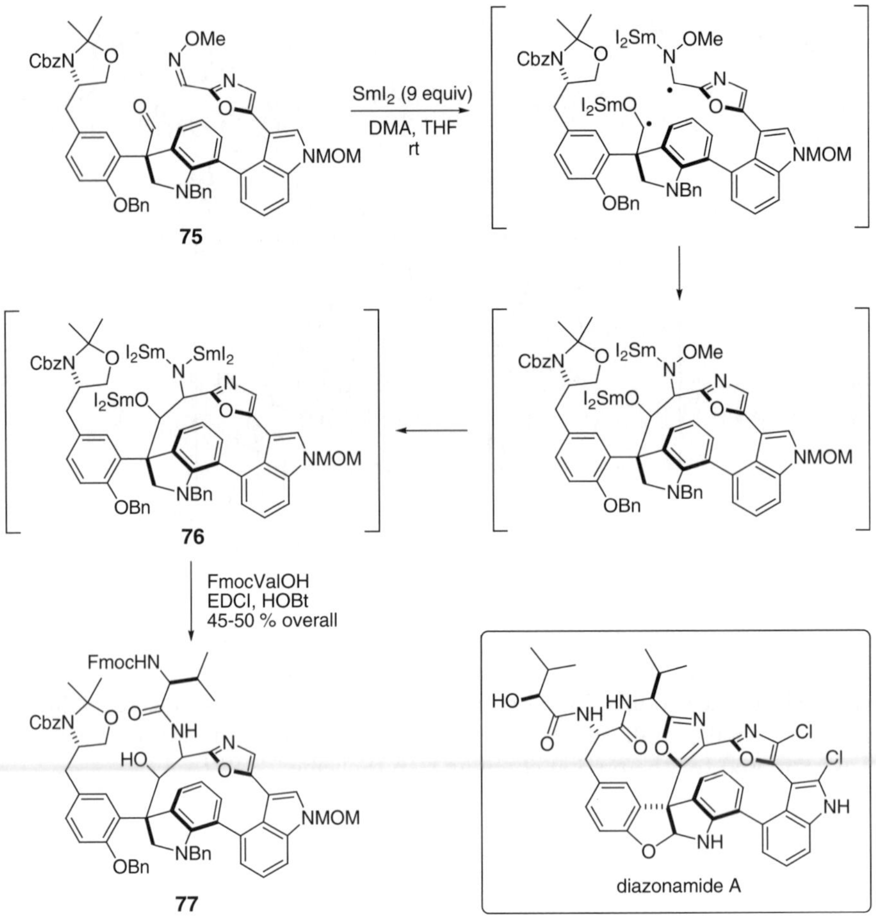

CbzN O N OMe N O NMOM O NBn OBn **75**

SmI_2 (9 equiv)
DMA, THF
rt

CbzN O I_2Sm N OMe N O I_2SmO NMOM NBn OBn

CbzN O I_2Sm N OMe N O I_2SmO NMOM NBn OBn

CbzN O I_2Sm N SmI_2 N O I_2SmO NMOM NBn OBn **76**

FmocValOH
EDCI, HOBt
45-50 % overall

FmocHN CbzN O O NH N O HO NMOM NBn OBn **77**

HO NH HN O O N O N O N Cl O Cl NH O NH

diazonamide A

Scheme 7.32

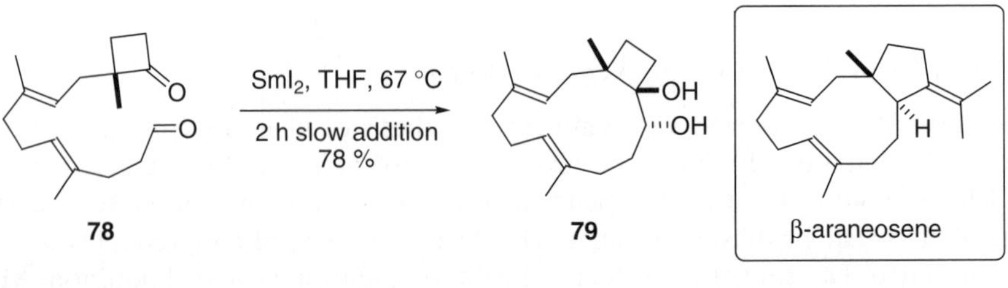

O O **78**

SmI_2, THF, 67 °C
2 h slow addition
78 %

OH OH **79**

H

β-araneosene

Scheme 7.33

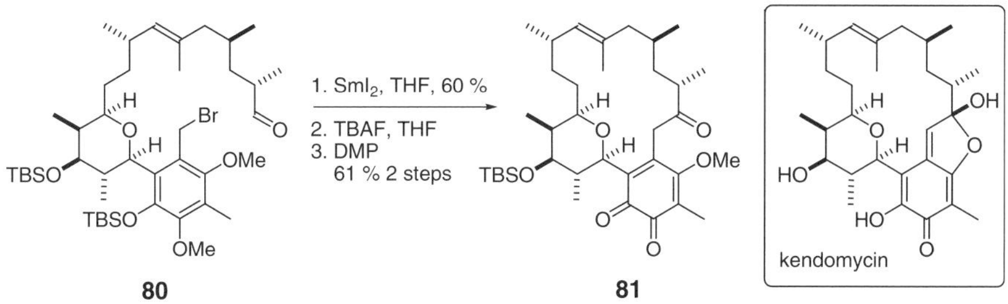

Scheme 7.34

7.4.8 Conclusion

This selective survey has showcased the use of SmI_2-mediated cyclisation reactions in natural product synthesis and pays testament to the power of SmI_2 as a reagent for synthesis. The full armoury of ring-forming reactions conducted by the reagent has been applied successfully in approaches to a wide range of cyclic systems in a wide range of targets.[55] Indeed, some of the most challenging targets of recent years, such as Taxol and diazonamide A, have been prepared using pivotal cyclisation steps mediated by the reagent. If natural product synthesis is a test for methods and reagents, then SmI_2 has passed and looks set to continue meeting the challenges we put before it.

7.5 Modifying Biomolecules Using SmI_2

7.5.1 Introduction

As discussed in previous chapters, SmI_2 promotes a wide variety of carbon–carbon bond-forming reactions while at the same time displaying a high degree of functional group tolerance. These properties make this low-valent lanthanide reagent particularly suitable for carrying out structural modifications on biomolecules, such as carbohydrates and peptides, with emphasis on the preparation of stable carbon-based mimics of interest for biomedical research and drug development programmes.

Performing selective chemical transformations and, in particular, carbon–carbon bond formation on such biomolecules has traditionally been challenging due to the many proximal functional groups found in such molecules (*i.e.* hydroxyl groups in sugars, amide bonds and side-chain functionalities in peptides). Nevertheless, in recent years it has been demonstrated that SmI_2 has a remarkable ability to promote a number of carbon–carbon bond-forming reactions on peptide and carbohydrate substrates. More specifically, SmI_2 has been used for the stereoselective synthesis of *C*-glycosides, side-chain introduction on to glycine residues of peptides, the ligation of small peptides and the direct synthesis of ketomethylene and hydroxyethylene isosteres. From a synthetic point of view, it is interesting that the high functional group tolerance of

SmI_2 allows intact carbohydrates and peptides to be used as building blocks in the synthesis of biomolecule mimics without resort to *de novo* synthesis.

7.5.2 Modifying Carbohydrates Using SmI_2

C-Glycosides represent a popular class of carbohydrate mimics that are characterised by the replacement of the interglycosidic oxygen atom by a methylene group. These analogues are not only resistant to enzymatic and chemical hydrolysis, but also display conformational properties around the glycosidic linkage that are similar to those of the parent *O*-glycosides. Over the last decade, SmI_2 has been found to be an excellent reagent for the preparation of *C*-glycosides because, unlike most other approaches to these compounds, the intact sugar can be used in the carbon–carbon bond-forming step without problems arising from the presence of the ring substituents. Two approaches involving the generation of a reactive anomeric intermediate have been used for the SmI_2-promoted synthesis of *C*-glycosides, the first involving a C1-carbon centred radical and the second involving a carbanion.

In 1994, Sinaÿ reported an extraordinary intramolecular delivery approach for the preparation of a *C*-disaccharide (Scheme 7.35).[74] The glycosyl sulfone **82** was tethered to the unsaturated sugar acceptor **83** via a silyl diether linkage. Upon slow addition of SmI_2 in benzene and HMPA, **84** underwent a 9-*endo* radical cyclisation to give **85**. Whereas control at the C1-position was achieved due to the anomeric effect and the large ring formed, 1,3-diaxial interactions guide the new C4-substituent of the reducing sugar into an equatorial position. Subsequent desilylation of the major C4-epimer **85** provided the disaccharide mimic **86** in an overall yield of 50% for the two steps (Scheme 7.35).[74]

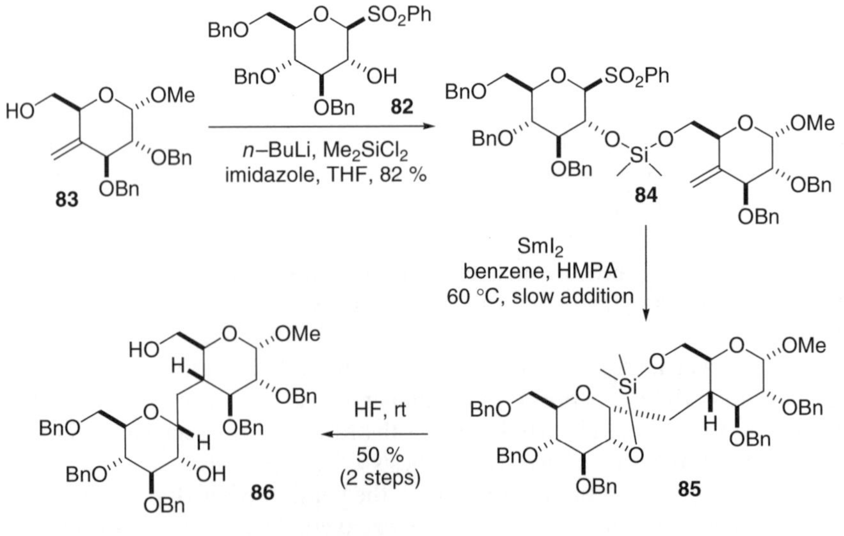

Scheme 7.35

Representative procedure. Dry benzene (45 ml) and HMPA (5 ml) were added to 1,2-diiodoethane (2.5 mmol) and Sm powder (3 mmol) under argon. The mixture was stirred at room temperature for 12 h under sonication to yield a 0.05 M solution of SmI_2. The substrate (1 equiv) was dissolved in toluene and warmed to 60 °C before syringe pump addition of the benzene–HMPA solution of SmI_2 (4.2 equiv) at a rate of $0.6\,ml\,h^{-1}$.

In an effort to avoid the use of HMPA as a cosolvent for the generation of anomeric radicals using SmI_2, Beau and Skrydstrup reported the use of the more reactive glycosyl pyridyl sulfones in conjunction with a radical acceptor tethered to the C2-OH for the stereoselective synthesis of α-*C*-glucosides and β-*C*-mannosides.[75,76] In these examples, the installation of the C1-alkyl group using a 5-*exo*-radical allows the stereochemistry of the C2-substituent to dictate the stereochemical outcome in the carbon–carbon bond-forming step at the anomeric centre. As exemplified in Scheme 7.36 with the pyridyl sulfone **87**, cyclisation occurs upon addition of SmI_2 in THF alone. Subsequent desilylation, double bond hydrogenation and peracetylation provided the methyl isomaltoside mimic **88** in an overall 48% yield (Scheme 7.36).[75,76]

Further work by Beau and Skrydstrup demonstrated the surprising stability of anomeric organosamarium species (generated from the reduction of glycosyl pyridyl sulfones with SmI_2) towards β-elimination of the protected C2-hydroxy substitutent.[77,78] Glycosyl samariums could even be trapped with alkyl ketones and aldehydes. These Barbier-type reactions (see Chapter 5, Section 5.4) generally proceed with good stereocontrol at the anomeric centre, leading to 1,2-*trans* selectivity (in contrast to the 1,2-*cis* selectivity observed in the SmI_2-mediated radical cyclisations shown in Schemes 7.35 and 7.36). For example, the reductive coupling of mannosyl pyridyl sulfones **89** with carbonyl compounds

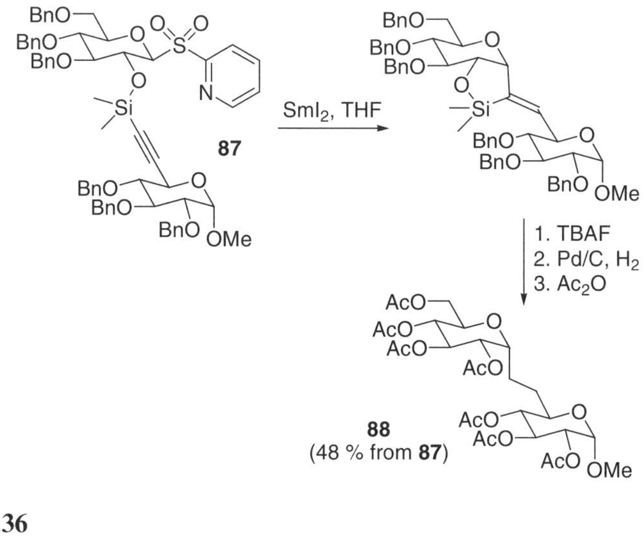

Scheme 7.36

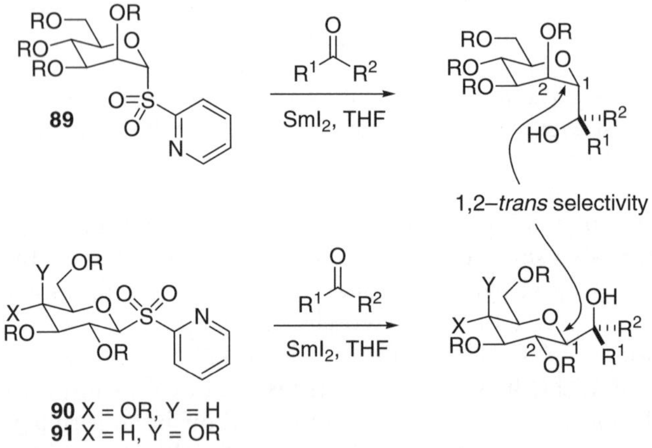

Scheme 7.37

led to α-*C*-mannoside formation, whereas the glucosyl and galactosyl sulfones **90** and **91** produced the corresponding β-*C*-glycosides (Scheme 7.37).[77,78]

Representative procedure. To a stirred solution of the sulfone (1 equiv) and ketone (1.4 equiv) at room temperature in THF was added a 0.1 M solution of SmI_2 (2.1 equiv). After 10 min, the reaction was quenched with saturated NH_4Cl. The aqueous layer was extracted with CH_2Cl_2 and the organic layers were washed with H_2O (×2). The combined organic layers were dried (Na_2SO_4) and the solvent was removed *in vacuo*.

The stereoselectivities observed at the C1-position of the sugar ring were explained by a consideration of the fate of the glycosyl anions formed from the reduction of the intermediate anomeric radical (Scheme 7.38). As these C1-radicals are thermodynamically more stable in the α-orientation, their reduction will lead to the kinetic axially oriented organosamarium species. Unfavourable overlap between the σ_{C1-Sm} and n_{O5} orbitals is relieved by either a conformational change of the ring to a skew boat, placing the C1- and C2-substituents in a diequatorial arrangement (as occurs in the manno series), or a configurational change (anomerisation) to the β-organosamarium species, as observed in the gluco and galacto series (Scheme 7.38). Surprisingly, β-elimination of the C2-substituent was not generally found to be the major pathway. In subsequent work, Beau demonstrated that the use of catalytic NiI_2 leads to a substantial increase in yield for these *C*-glycosylation reactions.[79]

In contrast, the Barbier reactions of 2-acetamido-2-deoxy sugars with ketones showed a preference for the formation of 1,2-*cis*-*C*-glycoside products (Scheme 7.39). This was attributed to the complexation of the C2-acetamido group to the glycosyl samarium, thereby reducing the configurational lability of the organometallic species.[78,80,81]

Further work from the groups of Linhardt and Beau has demonstrated the viability of performing SmI_2-mediated *C*-glycosylations using derivatives of

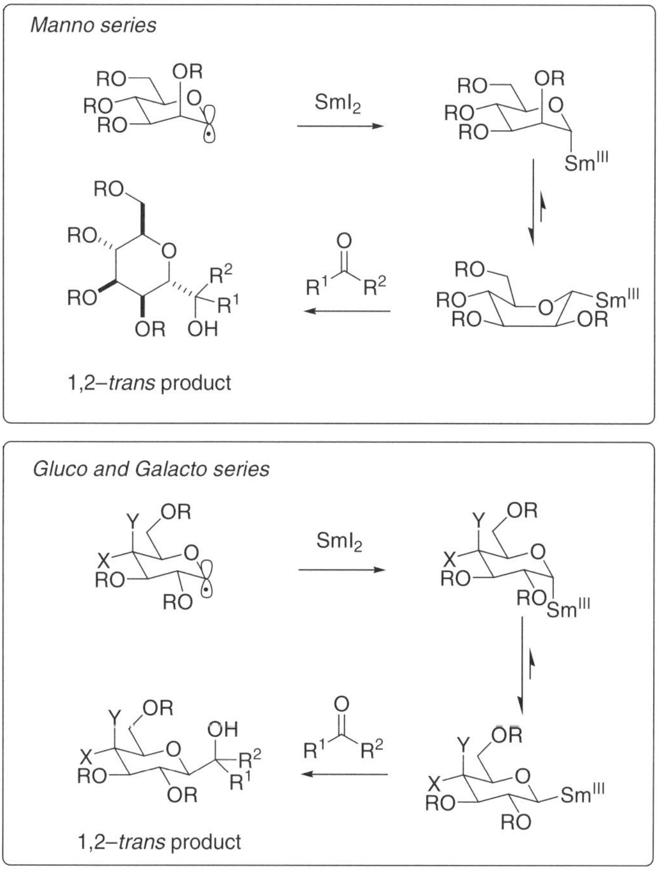
Manno series
SmI2
SmIII
1,2–trans product
Gluco and Galacto series
1,2–trans product

Scheme 7.38

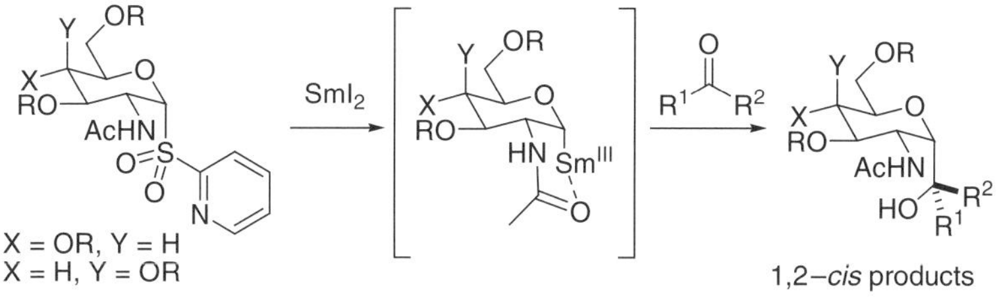
X = OR, Y = H
X = H, Y = OR
SmI2
1,2–cis products

Scheme 7.39

N-acetylneuraminic acid.[82–84] Most impressively, Beau has shown that simple peracetates of methyl *N*-acetylneuraminate such as **92** perform well in couplings with carbonyl compounds by a Reformatsky-type mechanism (see Chapter 5, Section 5.5) (Scheme 7.40).[84,85] SmI_2-mediated *C*-glycosylations of this type have also been carried out on a solid phase (see Section 7.3).

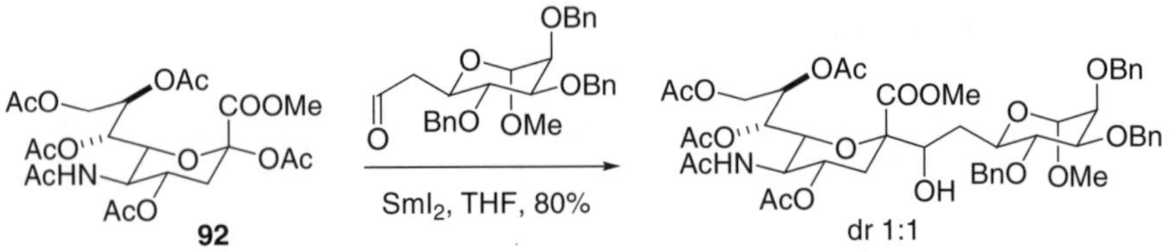

Scheme 7.40

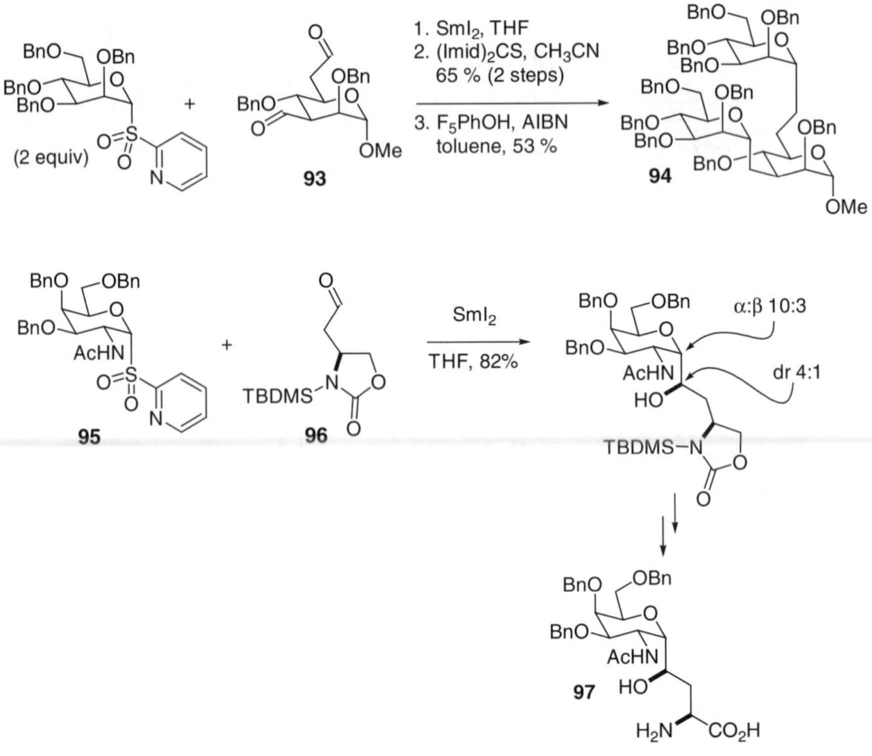

Scheme 7.41

Finally, two examples of the application of the above chemistry for the synthesis of a *C*-oligosaccharide and a *C*-glycosylated amino acid are illustrated in Scheme 7.41. Skrydstrup reported a highly convergent synthesis of the *C*-trisaccharide **94** that involves a double SmI_2-promoted *C*-glycosylation of the dialdehyde **93**, followed by a modified Barton–McCombie deoxygenation step (Scheme 7.41).[86] Furthermore, the Skrydstrup–Beau team disclosed a direct synthesis of the *C*-glycoside analogue **97** of a tumour-associated carbohydrate antigen (Tn), which exploits a SmI_2-mediated coupling of the pyridyl sulfone **95** with the aldehyde **96** (Scheme 7.41).[87]

7.5.3 Modifying Amino Acids and Peptides Using SmI_2

The ability of SmI_2 to promote carbon–carbon bond-forming reactions at the anomeric centre of carbohydrates, under mild coupling conditions, and the remarkable stability of the glycosyl samarium(III) species towards β-elimination (see Section 7.5.2) inspired Skrydstrup to examine the viability of analogous reactions on small peptide substrates for the introduction of non-natural carbinol side-chains to glycine residues.[88,89] Introduction of a reducible group on to the glycine residue was achieved in simple cases by bromination and subsequent nucleophilic displacement of the halide with 2-mercaptopyridine. However, a higher yielding and more tolerant protocol involved the oxidative fragmentation of a serine residue with $Pb(OAc)_4$ followed by substitution of the resultant acetate group with 2-mercaptopyridine. For example, the cyclic peptide **98** and the tripeptide **99** were converted to the corresponding sulfide and subsequent low-temperature coupling of the peptides to alkylaldehydes and ketones in the presence of SmI_2 and catalytic NiI_2 gave good yields of the modified peptide, although little stereocontrol at the α-carbon was achieved (Scheme 7.42). It is likely that the couplings involve Sm(III) enolate formation and aldol reaction with the carbonyl electrophile (see Chapter 5, Section 5.5). The low basicity of Sm(III) enolates limits proton abstraction from adjacent residues.

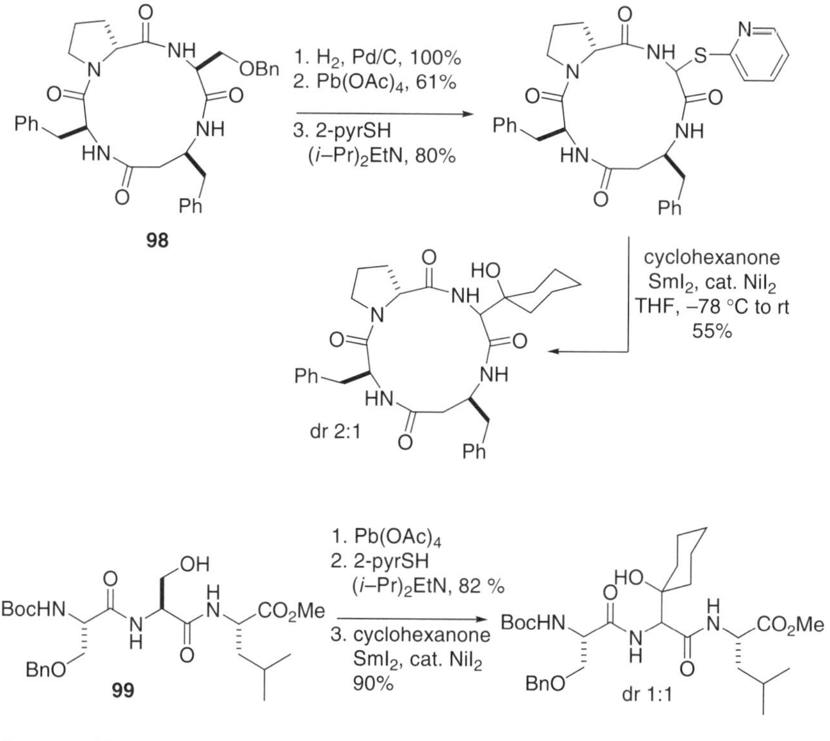

Scheme 7.42

Representative procedure. To a stirred solution of SmI_2 in THF (0.1 M, 1 equiv) was added a solution of NiI_2 in THF (0.01 M, 0.01 equiv) and the resultant mixture was stirred for 15 min before cooling to 0 °C. The cooled solution was then added dropwise to a stirred solution of the pyridyl sulfide (0.33 equiv) and cyclohexanone (1 equiv) in THF (1.2 M) at −78 °C. The reaction mixture was stirred for 10 min before warming to room temperature and quenching with saturated aqueous NH_4Cl. The aqueous layer was extracted with CH_2Cl_2 (×2). The combined organic phases were then washed with H_2O, filtered through Celite, dried ($MgSO_4$) and concentrated *in vacuo*. The crude product was purified by column chromatography.

In 2003, a new SmI_2-mediated carbon–carbon bond-forming reaction was reported by Skrydstrup for the direct synthesis of peptide mimics for evaluation as protease inhibitors.[90] For example, the low-temperature coupling of 4-thiopyridyl ester **100**, derived from Cbz-protected phenylalanine, with the dipeptide acrylamide **101** gave the peptide analogue **102** in a 61% yield (Scheme 7.43). Ketone **102** represents a ketomethylene isostere of the tetrapeptide Phe–Gly–Leu–Phe. Ketomethylene isosteres and the corresponding reduced analogues, hydroxyethylene isosteres, represent important and pharmaceutically relevant classes of protease inhibitors.[91,92]

The mechanism of these reactions was proposed to take place by initial electron transfer to the carbonyl group of the thioester, generating a ketyl-like radical anion such as **103** (Scheme 7.43). Subsequent radical addition to the electron-deficient alkene (acrylamide or acrylate), possibly guided by precomplexation to a Sm(III) metal ion, generates a new radical centre, which is reduced to the corresponding Sm(III) enolate by a second equivalent of SmI_2. Protonation of this enolate and hydrolysis of the hemithioacetal upon work-up then lead to the γ-ketoamide or ester.

Scheme 7.43

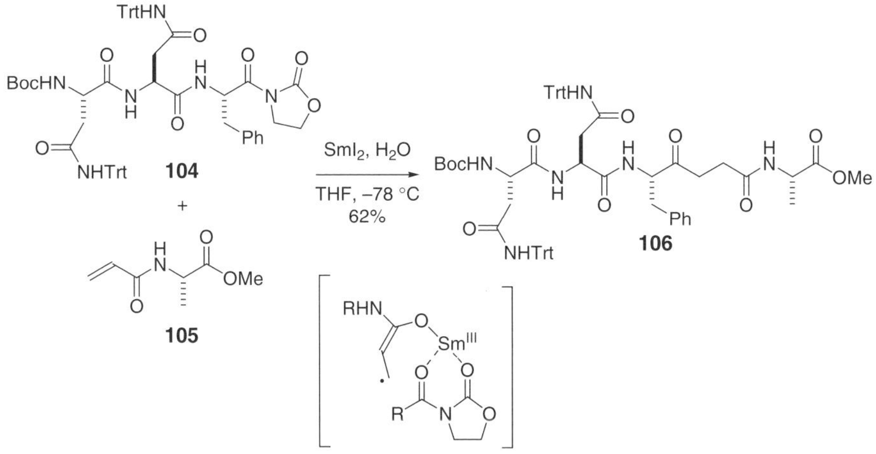

Scheme 7.44

Although the scope of the reaction using thioester-derived amino acids was found to be rather narrow with respect to the two coupling partners, further work revealed a more general protocol for accessing peptidyl ketones: SmI_2-mediated coupling of the readily accessible *N*-peptidyloxazolidinones with electron-deficient alkenes gave ketones in excellent yield.[93] For example, reductive coupling of the tripeptidyl derivative **104** and the acrylamide alanine **105** gave the desired peptidyl ketone **106** in 62% yield (Scheme 7.44).[94,95] In contrast to the coupling of thioesters with electron-deficient alkenes (see Scheme 7.43), the coupling of oxazolidinones with electron-deficient alkenes is thought to proceed by reduction of the acrylamide followed by addition of the resultant radical intermediate to the exocyclic carbonyl group of the *N*-acyloxazolidinone (Scheme 7.44).[96]

Representative procedure. To a stirred solution of oxazolidinone (1 equiv) and acrylamide (1.5 equiv) in THF was added H_2O (8 equiv). The reaction mixture was cooled to –78 °C under an argon atmosphere and a solution of SmI_2 in THF (0.1 M, 4 equiv) was added dropwise over 30 min and the mixture was stirred for 48 h. The reaction mixture was subsequently flushed with oxygen before the addition of saturated aqueous NH_4Cl, warmed to room temperature and then poured into 0.5 M HCl. The aqueous layer was extracted with EtOAc (×3) and the combined organic phases were washed with saturated aqueous $Na_2S_2O_3$, dried ($MgSO_4$) and concentrated *in vacuo*. The crude product was purified by column chromatography.

7.5.4 Conclusions

Although few groups have yet to exploit SmI_2 for the generation of carbon–carbon bonds within biomolecules such as carbohydrates, amino acids and peptides, the work highlighted in this section demonstrates once again the versatility of this reagent for promoting challenging anionic and radical reactions on highly functionalised substrates. There is no doubt that the reagent will be used in the future to perform selective chemistry on some of Nature's most important molecules.

References

1. P. W. Selwood, *J. Am. Chem. Soc.*, 1934, **56**, 2392.
2. A. Lebrun, J.-L. Namy and H. B. Kagan, *Tetrahedron Lett.*, 1993, **34**, 2311.
3. J. R. Fuchs, M. L. Mitchell, M. Shabangi and R. A. Flowers II, *Tetrahedron Lett.*, 1997, **38**, 8157.
4. F. Hélion, M.-L. Lannou and J.-L. Namy, *Tetrahedron Lett.*, 2003, **44**, 5507.
5. A. Dahlén and G. Hilmersson, *Eur. J. Inorg. Chem.*, 2004, 3020.
6. R. S. Miller, J. M. Sealy, J. R. Fuchs, M. Shabangi and R. A. Flowers II, *J. Am. Chem. Soc.*, 2000, **122**, 7718.
7. K. Rossmanith, *Monatsh. Chem.*, 1979, **110**, 109.
8. S. Matsukawa and Y. Hinakubo, *Org. Lett.*, 2003, **5**, 1221.
9. S. E. Reisman, J. M. Ready, M. M. Weiss, A. Hasuoka, M. Hirata, K. Tamaki, T. V. Ovaska, C. J. Smith and J. L. Wood, *J. Am. Chem. Soc.*, 2008, **130**, 2087.
10. S.-i. Fukuzawa, T. Tsuchimoto and T. Kanai, *Chem. Lett.*, 1994, 1981.
11. T. Hanamoto, Y. Sugimoto, A. Sugino and J. Inanaga, *Synlett*, 1994, 377.
12. K. Mashima, T. Oshiki and K. Tani, *J. Org. Chem.*, 1998, **63**, 7114.
13. J. Collin, N. Giuseppone, F. Machrouhi, J.-L. Namy and F. Nief, *Tetrahedron Lett.*, 1999, **40**, 3161.
14. S.-i. Fukuzawa, K. Mutoh, T. Tsuchimoto and T. Hiyama, *J. Org. Chem.*, 1996, **61**, 5400.
15. S.-i. Fukuzawa, Y. Yahara, A. Kamiyama, M. Hara and S. Kikuchi, *Org. Lett.*, 2005, **7**, 5809.
16. J. L. Namy, J. Collin, J. Zhang and H. B. Kagan, *J. Organomet. Chem.*, 1987, **328**, 81.
17. W. J. Evans, J. W. Grate, H. W. Choi, I. Bloom, W. E. Hunter and J. L. Atwood, *J. Am. Chem. Soc.*, 1985, **107**, 941.
18. C. Bied, J. Collin and H. B. Kagan, *Tetrahedron*, 1992, **48**, 3877.
19. W. J. Evans, I. Bloom, W. E. Hunter and J. L. Atwood, *J. Am. Chem. Soc.*, 1981, **103**, 6507.
20. R. A. Flowers II and E. Prasad, in *Handbook on the Physics and Chemistry of Rare Earths*, ed. K. A. Gschneidner Jr, J.-C. G. Bunzli and V. K. Pecharsky, Elsevier, Amsterdam, 2006, **Vol. 36**, p. 393.
21. W. J. Evans, *Coord. Chem. Rev.*, 2000, **206**, 263.
22. W. J. Evans, *Polyhedron*, 1987, **6**, 803.

23. Z. Hou, A. Fujita, Y. Zhang, T. Miyano, H. Yamazaki and Y. Wakatsuki, *J. Am. Chem. Soc.*, 1998, **120**, 754.
24. E. K. Brady, D. L. Clark, D. W. Keogh, B. L. Scott and J. G. Watkin, *J. Am. Chem. Soc.*, 2002, **124**, 7007.
25. M. Kim, B. W. Knettle, A. Dahlén, G. Hilmersson and R. A. Flowers II, *Tetrahedron*, 2003, **59**, 10397.
26. N. Marques, A. Sella and J. Takats, *Chem. Rev.*, 2002, **102**, 2137.
27. R. Nomura, T. Matsuno and T. Endo, *J. Am. Chem. Soc.*, 1996, **118**, 11666.
28. H. C. Aspinall, N. Greeves and C. Valla, *Org. Lett.*, 2005, **7**, 1919.
29. F. Orsini and E. M. Lucci, *Tetrahedron Lett.*, 2005, **46**, 1909.
30. E. J. Corey and G. Z. Zheng, *Tetrahedron Lett.*, 1997, **38**, 2045.
31. F. Hélion and J.-L. Namy, *J. Org. Chem.*, 1999, **64**, 2944.
32. M.-I. Lannou, F. Hélion and J.-L. Namy, *Tetrahedron*, 2003, **59**, 10551.
33. C. Magnier-Bouvier, I. Reboule, R. Gil and J. Collin, *Synlett*, 2008, 1211.
34. J. M. Concellón, P. L. Bernad and E. Bardales, *Chem. Eur. J.*, 2004, **10**, 2445.
35. F. Xu, J. Sun and Q. Shen, *Tetrahedron Lett.*, 2002, **43**, 1867.
36. F. Xu, Y. Luo, M. Deng and Q. Shen, *Eur. J. Org. Chem.*, 2003, 4728.
37. J. Collin, N. Jaber and M. I. Lannou, *Tetrahedron Lett.*, 2001, **42**, 7405.
38. Y. H. Kim and H. S. Park, *Synlett*, 1998, 261.
39. D. A. Evans and A. H. Hoveyda, *J. Am. Chem. Soc.*, 1990, **112**, 6447.
40. L. A. Sloan and D. J. Procter, *Chem. Soc. Rev.*, 2006, **35**, 1215.
41. X. Du and R. W. Armstrong, *J. Org. Chem.*, 1997, **62**, 5678.
42. X. Du and R. W. Armstrong, *Tetrahedron Lett.*, 1998, **39**, 2281.
43. S. N. Baytas, Q. Wang, N. A. Karst, J. S. Dordick and R. J. Linhardt, *J. Org. Chem.*, 2004, **69**, 6900.
44. N. J. Kerrigan, P. C. Hutchison, T. D. Heightman and D. J. Procter, *Chem. Commun.*, 2003, 1402.
45. N. J. Kerrigan, P. C. Hutchison, T. D. Heightman and D. J. Procter, *Org. Biomol. Chem.*, 2004, 2476.
46. J. C. Vogel, R. Butler and D. J. Procter, *Tetrahedron*, 2008, **64**, 11876.
47. R. M. Myers, S. P. Langston, S. P. Conway and C. Abell, *Org. Lett.*, 2000, **2**, 1349.
48. M. M. Meloni and M. Taddei, *Org. Lett.*, 2001, **3**, 337.
49. F. M. McKerlie, D. J. Procter and G. Wynne, *Chem. Commun.*, 2002, 584.
50. L. A. McAllister, K. L. Turner, S. Brand, M. Stefaniak and D. J. Procter, *J. Org. Chem.*, 2006, **71**, 6497.
51. L. A. McAllister, R. A. McCormick, S. Brand and D. J. Procter, *Angew. Chem. Int. Ed.*, 2005, **44**, 452.
52. R. A. McCormick, K. M. James, S. Brand, N. Willetts and D. J. Procter, *Chem. Eur J.*, 2007, **13**, 1032.
53. K. M. James, N. Willetts and D. J. Procter, *Org. Lett.*, 2008, **10**, 1203.
54. M. Miller, J. C. Vogel, W. Tsang, A. Merrit and D. J. Procter, *Org. Biomol. Chem.*, 2009, 589.

55. D. J. Edmonds, D. Johnston and D. J. Procter, *Chem. Rev.*, 2004, **104**, 3371.
56. E. J. Corey and Y. J. Wu, *J. Am. Chem. Soc.*, 1993, **115**, 8871.
57. D. J. Edmonds, K. W. Muir and D. J. Procter, *J. Org. Chem.*, 2003, **68**, 3190.
58. T. M. Baker, D. J. Edmonds, D. Hamilton, C. J. O'Brien and D. J. Procter, *Angew. Chem. Int. Ed.*, 2008, **47**, 5631.
59. T. L. Fevig, R. L. Elliot and D. P. Curran, *J. Am. Chem. Soc.*, 1988, **110**, 5064.
60. T. Kan, S. Hosokawa, S. Nara, M. Oikawa, S. Ito, F. Matsuda and H. Shirahama, *J. Org. Chem.*, 1994, **59**, 5532.
61. P. Chen, J. Wang, K. Liu and C. Li, *J. Org. Chem.*, 2008, **73**, 339.
62. K. Kawamura, H. Hinou, G. Matsuo and T. Nakata, *Tetrahedron Lett.*, 2003, **44**, 5259.
63. M. Kitamura, K. Ohmori, T. Kawase and K. Suzuki, *Angew. Chem. Int. Ed.*, 1999, **38**, 1229.
64. J. Park, B. Kim, H. Kim, S. Kim and D. Kim, *Angew. Chem. Int. Ed.*, 2007, **46**, 4726.
65. H. Fukui and I. Shiina, *Org. Lett.*, 2008, **10**, 3153.
66. H. Miyabe, M. Torieda, K. Inoue, K. Tajiri, T. Kiguchi and T. Naito, *J. Org. Chem.*, 1998, **63**, 4397.
67. D. Riber, R. Hazell and T. Skrydstrup, *J. Org. Chem.*, 2000, **65**, 5382.
68. T. Mukaiyama, I. Shiina, H. Iwadare, M. Saitoh, T. Nishimura, N. Ohkawa, H. Sakoh, K. Nishimura, Y.-i. Tani, M. Hasegawa, K. Yamada and K. Saitoh, *Chem. Eur. J.*, 1999, **5**, 121.
69. G. A. Molander, K. M. George and L. G. Monovich, *J. Org. Chem.*, 2003, **68**, 9533.
70. M. Inoue, M. Sasaki and K. Tachibana, *Angew. Chem. Int. Ed.*, 1998, **37**, 965.
71. K. C. Nicolaou, P. B. Rao, J. Hao, M. V. Reddy, G. Rassias, X. Huang, D. Y.-K. Chen and S. A. Snyder, *Angew. Chem. Int. Ed.*, 2003, **42**, 1753.
72. J. Kingsbury and E. J. Corey, *J. Org. Chem.*, 2005, **127**, 13813.
73. J. T. Lowe and J. S. Panek, *Org. Lett.*, 2008, **10**, 3813.
74. A. Chénedé, E. Perrin, E. D. Rekaï and P. Sinaÿ, *Synlett*, 1994, 420.
75. D. Mazéas, T. Skrydstrup, O. Doumeix and J.-M. Beau, *Angew. Chem. Int. Ed. Engl.*, 1994, **33**, 1383.
76. T. Skrydstrup, J.-M. Beau, M. Elmouchir, D. Mazeas, C. Riche and A. Chiaroni, *Chem. Eur. J.*, 1997, **3**, 1342.
77. D. Mazéas, T. Skrydstrup and J.-M. Beau, *Angew. Chem. Int. Ed. Engl.*, 1995, **34**, 909.
78. T. Skrydstrup, O. Jarreton, D. Mazéas, D. Urban and J.-M. Beau, *Chem. Eur. J.*, 1998, **4**, 655.
79. N. Miquel, G. Doisneau and J.-M. Beau, *Angew. Chem. Int. Ed.*, 2000, **39**, 4111.
80. D. Urban, T. Skrydstrup, C. Riche, A. Chiaroni and J.-M. Beau, *Chem. Commun.*, 1996, 1883.

81. L. Andersen, L. M. Mikkelsen, J.-M. Beau and T. Skrydstrup, *Synlett*, 1998, 1393.
82. I. R. Vlahov, P. I. Vlahova and R. J. Linhardt, *J. Am. Chem. Soc.*, 1997, **119**, 1480.
83. H. G. Bazin, Y. Du, T. Polat and R. J. Linhardt, *J. Org. Chem.*, 1999, **64**, 7254.
84. A. Malapelle, Z. Abdallah, G. Doisneau and J.-M. Beau, *Angew. Chem. Int. Ed.*, 2006, **43**, 6016.
85. A. Malapelle, A. Coslovi, G. Doisneau and J.-M. Beau, *Eur. J. Org. Chem.*, 2007, 3145.
86. L. M. Mikkelsen, S. L. Krintel, J. Jiménez-Barbero and T. Skrydstrup, *J. Org. Chem.*, 2002, **67**, 6287.
87. D. Urban, T. Skrydstrup and J.-M. Beau, *Chem. Commun.*, 1998, 955.
88. M. Ricci, P. Blakskjær and T. Skrydstrup, *J. Am. Chem. Soc.*, 2000, **122**, 12414.
89. P. Blakskjær, A. Gavrila, L. Andersen and T. Skrydstrup, *Tetrahedron Lett.*, 2004, **45**, 9091.
90. P. Blakskjær, B. Høj, D. Riber and T. Skrydstrup, *J. Am. Chem. Soc.*, 2003, **125**, 4030.
91. L. M. Mikkelsen, C. M. Jensen, B. Høj, P. Blakskjær and T. Skrydstrup, *Tetrahedron*, 2003, **59**, 10541.
92. K. B. Lindsay and T. Skrydstrup, *J. Org. Chem.*, 2006, **71**, 4766.
93. C. M. Jensen, K. B. Lindsay, R. H. Taaning, J. Karaffa, A. M. Hansen and T. Skrydstrup, *J. Am. Chem. Soc.*, 2005, **127**, 6544.
94. T. Mittag, K. L. Christensen, K. B. Lindsay, N. C. Nielsen and T. Skrydstrup, *J. Org. Chem.*, 2008, **73**, 1088.
95. J. Karaffa, K. B. Lindsay and T. Skrydstrup, *J. Org. Chem.*, 2006, **71**, 8219.
96. A. M. Hansen, K. B. Lindsay, P. K. Sudhadevi Antharjanam, J. Karaffa, K. Daasbjerg, R. A. Flowers II and T. Skrydstrup, *J. Am. Chem. Soc.*, 2006, **128**, 9616.

Subject Index